Digital Techniques in Broadcasting Transmission

Digital Techniques in Broadcasting Transmission

Robin Blair

Focal Press
Taylor & Francis Group

NEW YORK AND LONDON

First published 2002 by Focal Press
This edition published 2013
by Focal Press
70 Blanchard Road, Suite 402, Burlington, MA 01803

Simultaneously published in the UK
By Focal Press
2 Park Square, Milton Park, Abingdon, Oxon OX14 4RN

Focal Press is an imprint of the Taylor & Francis Group, an informa business

Library of Congress Cataloging-in-Publication Data
Blair, Robin.
    Digital techniques in broadcasting transmission / Robin Blair.
        p. cm.
    Includes index.
    ISBN 0-240-80508-9 (alk. paper)
    1.   Digital television. 2.   Television—Transmitters and transmission. 3.   Digital audio broadcasting. I.   Title.

    TK6678.B53 2002
    621.388—dc21

                                                                                                                2002018829

British Library Cataloguing-in-Publication Data
A catalogue record for this book is available from the British Library.

ISBN: 978-0-240-80508-5 (pbk)
ISBN: 978-0-080-50453-7 (ebk)

Contents

Preface

Digital techniques have been used in TV and sound studios for years, with quite spectacular success. Now the time has come to introduce digital techniques into the transmission chain. However, few people engaged in that side of broadcasting have had much exposure to the basic principles and techniques that underlie digital transmission systems. To fill that gap, this book is intended to be a primer on the theory of digital transmission with an emphasis on its application to broadcasting.

The book is structured around the approach of discussing the principles and then showing how they are being applied in practice in the systems now emerging. The definition of a transmission system is fairly broad. In this context, video and audio compression systems reduce the bit rates required to well-below studio standards and make their transmission practical within the bandwidths available for broadcasting. In the author's opinion, compression is undoubtedly part of the transmission system, and an extremely interesting one at that. Hence, they are treated in some detail, especially the famous MPEG video compression standards whose influence is spreading into myriad applications as well as broadcasting.

The practical systems described in the book encompass the ATSC 8-VSB TV Standard developed in the United States, the European Digital Video Broadcasting (DVB) standards, Eureka 147, the new Digital Radio Mondiale (DRM) digital audio broadcasting systems, and some satellite applications. The treatment focuses on systems that are, or are likely to become, recognized world standards. There is a short history of these in Chapter 1.

Several of the chapters are aimed at the practical aspects of engineering digital systems in the field. The engineering environment is similar to, but not the same as, what we have experienced in analog transmission. Some of what was regarded as vitally important in analog transmission now hardly matters at all, but naturally the digital medium does introduce much that is new. It is the author's opinion, however, that once the principles are understood, station engineering is probably simpler with digital than it has been with analog.

It will be a good many years before the analog channels disappear completely, and in the meantime analog and digital will have to live together. Many existing operators will find themselves having to transmit their original analog service and the new digital services in parallel. Hence, the book explores what is required in these situations. It contains background material on sharing antennas and channel combining, and particularly adjacent channel combining. These will be very useful techniques during the transition period to full digital.

The book generally avoids the use of mathematics except at a basic level. Some equations are included where they may be a helpful introduction to readers who may wish to pursue a subject at greater depth in other texts. Nevertheless, one's appreciation of some of the finer theoretical details is much enhanced by an understanding of the Fourier transform. A short appendix explains the use of this in a way that, it is hoped, will be readily understood and appreciated by non-mathematical readers.

Chapter 1

Introduction

Why Digital?

There cannot be many human activities left on this earth where digital techniques have not taken over from older analog methods and vastly improved the end results. Thus we see applications from telephony to the control of automobile engines that have taken the digital route. Probably the one that is most obvious to the general public is the way the compact disk has almost totally replaced the vinyl record in a remarkably short time. And now even the compact disk is being replaced by the DVD. As an older engineer I can remember another example—the way the electronic calculator displaced the slide rule that had been an icon of our profession for centuries.

Yet broadcasting systems have remained staunchly analog, both for radio and TV. This is almost certainly because very good compression algorithms and the means to implement them have only recently become available. Compression is the process by which the digital bit rate required to transmit high-quality audio or video can be reduced to some value compatible with the channel bandwidths available to broadcasting services. A compact disk player typically utilizes a bit rate of about 1.5 Mbits/s, and in an uncompressed form video transmission would require over 200 Mbits/s. (Mbits/s is short for megabits per second. Readers unfamiliar with such terms are referred to the glossary at the end of the book.) Later in the book we will see that transmitting 200 Mbits/s would require a bandwidth of the order of 50 MHz. Bandwidth of that order is simply not available for the radiated services and would confine even cable systems to carrying only two or three services.

Attempts to reduce the bandwidth required for video transmission began not long after World War II, and some of the techniques described herein were known in the 1960s. However, their implementation had to await that other digital revolution— the development of large-scale integration chips with sufficient computation power to run the algorithms in real time. The advances since have been dramatic. Most major countries have announced their intention to have digital TV and radio services operating in the immediate future, and many have them on air already. For this they have relied on recently developed international standards. The standards have had a short but intense history of development. Let us look briefly at how the major ones have come into being.

Digital TV in the United States

During the 1980s, quality was becoming something of an issue for conventional broadcasters. Typical domestic video displays compared very poorly with what people were seeing in the cinema, and the emerging compact disks were seen as a real threat to music programs on radio. There were intense efforts to develop high definition analog TV (HDTV), but all required a relatively large bandwidth and, it seemed, could be delivered only by cable. The broadcasters naturally saw this as a threat to their businesses, and in 1987 a large group in the United States persuaded the Federal Communications Commission (FCC) to set up the Advisory Committee on Advanced Television Systems (ACATS).

ACATS set itself the task of testing candidate HDTV systems and selecting a North American Standard by June 1990. All the systems initially submitted to it were analog, but one digital system entered the fray just before the target cut-off date. ACATS sought an extension and by March 1992 had four proposals for digital systems and only two for analog, one of which was subsequently withdrawn. After intensive testing ACATS came down firmly on the side of digital transmission and recommended that the four proponent groups be given time to refine their systems. Seven of the major organizations involved then elected to form a cooperative known as the Grand Alliance, with the aim of combining the best features of the four systems to create the best possible standard. At this time also there was an intensive worldwide effort to develop a universal video and audio compression system. The result became the famous MPEG standards (from the semi-official title of the committee, Moving Picture Experts Group). The Grand Alliance adopted many features of the MPEG standards for incorporation into its system.

Two and a half years of intensive testing and refinement followed. The Advanced Television System Committee (ATSC) Standard was agreed in November 1995, and in December 1996 the FCC endorsed this as the digital TV (DTV) standard for the United States.

The DVB Project

There had been a number of attempts in Europe to devise an enhanced TV standard using analog techniques. One of the more famous was the MAC (multiple analog components) scheme. This compressed the components of the video signal in time and transmitted them in sequence, rather than in parallel, as happens in PAL or NTSC, and thus avoided many of the artifacts caused by mixing between the video components. It looked so much like the system of the future that at one time the European Commission legislated to make it compulsory in all new TV sets.

In 1990, however, General Instruments in the United States had demonstrated its Digicipher video compression system, which showed real promise for truly useful results in digital transmission. It was arguably this system that influenced ACATS away from the competing advanced analog proposals, and in addition its concepts formed the basis of the initial work by MPEG. That committee released its first standard in 1993. Although that first standard was aimed more at video storage and computer applications than broadcasting, its success made it obvious that the future of advanced television systems lay with digital. In 1993, the DVB Project was born with the expressed aim of developing world standards for universal digital television broadcasting.

DVB is an alliance of manufacturers, broadcasters, government agencies, and research institutions that has come to have over 300 members. Its aim is to develop digital TV broadcasting signal formats for delivery by terrestrial transmitters, cable systems, and satellites. There is a very high degree of commonality among all of these formats so that consumers may very easily receive any or all of them on the one domestic receiver. The European Telecommunications Standards Institute (ETSI) has adopted the formats developed by DVB as European Standards.

4

The Eureka 147 Project

Like DVB, Eureka 147 is a collaborative project being undertaken by a consortium of organizations with the aim of developing a world standard for digital audio broadcasting (DAB). It has over 40 members in total, mostly drawn from Europe but with significant contributions being made from other countries, notably Canada from the early days and lately Japan. The consortium was first formed in 1987. It has developed a multichannel sound broadcasting system with the compression based on the work of MPEG. The system was adopted as an ETSI Standard in early 1995, and the first commercial services went to air in Stockholm and London at the end of that year.

IBOC and IBAC DAB

As a multichannel system, Eureka 147 does not particularly suit broadcasters in the United States, where most operators are licensed to transmit only one program. It is more suited to the situation where big public or private organizations provide integrated suites of programs, such as those that exist in many European countries and Canada. In addition, its implementation almost inescapably requires that large new blocks of spectrum be made available for broadcasting, or that the existing spectrum be extensively reallocated amongst new and existing users. There are a number of countries, including the United States in particular, where these conditions are inappropriate to the commercial and legislative environments. Hence, developers in the United States have been working on systems known as "in band on channel" (IBOC) or "in band adjacent channel" (IBAC). These are specifically designed to permit an existing broadcaster to transmit a single digital service either within his existing channel or the one adjacent to it, which will almost always be vacant in his allocated service area.

5

Many of the same considerations that apply in the United States to terrestrial broadcasting also control the requirements for digital transmission in international HF broadcasting. Hence, many of the organizations involved in this activity had also been working to develop a narrow band IBOC system. In 1998, following a pattern set by the formation of the Grand Alliance, DVB, and MPEG, these organizations came together in a formal collaboration known as Digital Radio Mondiale, or DRM. USA Digital Radio, itself a collaborative group in the United States, was one of the earlier members of DRM. DRM was successful in having its proposed digital system adopted as a world standard in April 2001, for broadcasting at frequencies below 30 MHz.

At this time there are no accepted standards for IBOC systems in the current FM bands. However, USA Digital Radio and Lucent Digital Radio have come together to form iBiquity Digital Corporation with the express intention of promoting such a standard.

The MPEG Standards

No introductory chapter on digital broadcasting would be complete without some description of the MPEG Standards, since they underlie almost all current applications—and are becoming ubiquitous on the Internet as well. More formally they are known as the ISO/MPEG standards, where ISO refers to the International Standards Organization. As the name "Moving Picture Expert Group" signifies, the standards are essentially concerned with the digital coding and transmission of video and associated audio. The standards encompass three major elements—video compression, audio compression, and the structuring of the digital signal into a "transport stream." Later chapters in this book describe each of these elements in some detail.

MPEG-1 was the first standard and was aimed at low bit rate applications of relatively low quality material, e.g., computer storage. MPEG-2 came next and is the one most often found in broadcasting applications. MPEG-3 was intended to encompass very high quality and demanding applications, but many of its features came to be included in MPEG-2, and MPEG-3 essentially disappeared. MPEG-4 is the latest, and extends the MPEG reach into very creative multimedia applications.

The common usage of MPEG nomenclature can be somewhat confusing, in that the audio compression and transmission methods encompassed within the standards are termed "layers" and are not always prefixed by the label of the main standard. The audio components in the standards are formally MPEG-1, MPEG-2 Layer 1, MPEG-2 Layer 2, MPEG-2 Layer 3, and the very latest MPEG-2 AAC (advanced audio coding). However, one will often see audio coding referred to as MPEG 3, or lately MP3 on the Internet, where this really refers to MPEG-2 Layer 3.

The MPEG standards have found almost universal worldwide application. A major exception is in the audio component of the ACTS TV system in the United States, which employs a proprietary system developed by Dolby Laboratories. There are, however, many similarities between the Dolby and MPEG approaches. A second exception appears to be within the iBiquity IBOC system. It is described in the literature as employing a Perceptual Audio Coding (PAC) system, but this seems also to be closely related to MPEG-2 AAC.

A New Engineering Paradigm

Thomas Kuhn in his book *The Structure of Scientific Revolutions* first adopted the word *paradigm* to describe the set of theories, beliefs, and tests, adopted by the practitioners in some field of science from time to time. Analog broadcasting transmission has certainly had its own paradigm. For example, we have seen the importance placed on a myriad of waveform tests in TV testing and the great deal of attention given to impedance matching in antenna and transmission line systems in station design.

The digital era will require a new paradigm. We will show in this book that waveform testing is no longer appropriate and is displaced by reference to constellation diagrams, which bear no discernable relation to picture content. The reader will see that the digital systems are specifically designed to resist echoes far worse than those that arise from small mismatches in antenna systems. Hence, very strict impedance matching may become a thing of the past. At the same time, the transmission engineer will have to become familiar with some new requirements, for example, describing voltage and power in statistical terms rather than as defined by standard waveforms.

However, there is nothing really magical about digital transmission, and anyone with some knowledge of broadcasting will be able to understand the principles easily. At worst, one may have to come to terms with the Fourier transform to switch one's thinking between the time and the frequency domains. Even this, however, is a concept with which many broadcasting engineers are familiar, particularly those who have worked in television.

That said, there is one particular difficulty that engineers engaged in television transmission will have to face up to in the next few years. This is the mixed analog and digital environment. It is unlikely that the general public will accept the closing down of the existing analog services in the near future. Some governments have foreshadowed regulations intended to bring this about, but it will be interesting to see how effective that may prove to be. Consequently, the analog and digital transmissions will have to coexist for some time. Given the notorious scarcity of spectrum, and the environmental issues surrounding the establishment of new transmission infrastructure, we need to develop engineering systems catering for both analog and digital transmissions simultaneously. That is no simple task.

This book lays out systematically the principles that lie behind the current standards for digital broadcasting. It also discusses the mixed digital and analog environment to outline techniques that have been learned in the early years of introducing digital. It will be particularly useful to those who have a background or interest in the current analog transmission systems, and now wish to expand their knowledge into this new era.

Chapter 2

Sending by Numbers: Basic Digital Transmission

This chapter describes the essential concepts of digital transmission. It is a fortunate fact of life that all the essential theories and concepts can be derived from an elementary baseband model—sometimes referred to as a prototype system. Hence, the chapter will first describe this model and use it to show how and why the fundamental characteristics of digital transmission arise. Hopefully, the reader will find that these background processes are conceptually quite simple and almost obvious in most cases.

To describe digital transmission as simple may elicit surprise in those who may be having an initial exposure to system standards or to even mildly advanced descriptions of the systems in operation. The standards, in particular, can seem very much opaque to the uninitiated, as they always assume a great deal of background in the reader. That complexity, however, does not arise from the base principles of digital transmission, but rather from what might be called the administrative overhead. In the administrative overhead we can include such things as the need to combine digital streams from many sources, to include forward error correction, to combat propagation effects, and the like. It is essential to take these important topics one at a time to understand how complete systems operate, and to this end each is accorded a separate chapter in this book. For now, however, let us concentrate on the essential principles of simply transmitting a stream of digits from one place to another.

The prototype baseband system to be introduced in this chapter is the basis of many systems spanning a much broader field than the broadcasting applications on which we focus in this book. It is behind practically all wireline, cable, radio, and optical fiber applications. One finds its principles applied in fields as diverse as mobile telephony and computer modems. The essential differences appear in two areas—the administrative overheads mentioned previously, and also in the modulation systems used to impress the baseband signals onto radio frequency carriers.

The modulation systems used in broadcasting applications are outlined in this chapter and described in more detail in the other chapters that deal with particular systems and standards. The modulation systems used in most other applications outside broadcasting do not differ all that much in principle. There has been a trend, over years of development, for the modulation systems to incline towards those regarded as linear, mainly amplitude modulation in its various guises including Vestigial Sideband (VSB). Non-linear systems, like frequency modulation, have largely disappeared. This, at least, has made the modulation process easy to understand and to analyze. Unfortunately, the terms "phase modulation" or "phase shift keying" have persisted in the literature, but as we shall show, these are really misnomers and describe quite straightforward methods of applying amplitude modulation.

At the end of this chapter, therefore, the reader should have a good working knowledge of baseband digital transmission and of modern modulation systems, applicable not only to broadcasting but to most other fields as well.

Elementary Baseband Transmission

Digital transmission is all about sending and receiving symbols. The symbols belong to a defined and limited set, and only those that belong to that set may be sent. Each symbol represents a very specific value, usually a number or letter. It is the task of the receiver to decide which symbol has been sent and to assign the correct value to it. The analogy with human spoken communication is compelling. There, the speaker and listener share a vocabulary of a finite number of words, albeit possibly a very large number. The task of the listener is to decide which word has been spoken, amongst the likely possibilities. Thus, even at a noisy cocktail party, say, the listener may hear a word imperfectly but still be able to decide which word it is. If he or she generally makes correct decisions, then good understanding is possible even through the noisy communication channel that one inevitably finds at cocktail parties. If we equate "word" in this context to "symbol" in digital transmission we have a neat analogy to what the latter is required to achieve.

The simplest digital transmission one can imagine would be for transmitting binary digits. These can take only the values 1 and 0, and so only two different symbols are necessary, and a good choice might be a positive voltage pulse for the value 1 and a negative voltage pulse for 0. An elementary diagram for this system and the transmission waveform for an arbitrary stream of data is shown in Figure 2.1. In this, the receiver samples the voltage on the line at appropriate times, the sampling instances, to see if that voltage is positive or negative.

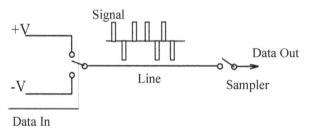

Figure 2.1: Elementary Digital Transmission System

Pulse Shaping and Bandwidth

Some things are immediately obvious. The more the pulses are narrowed and squeezed together in time, the greater the data transmission rate. The limitation here however is in the bandwidth that can be afforded to the system. There is a well-known inverse relationship between the width of a voltage pulse and the bandwidth required for its transmission, and the pulse will be badly distorted if that bandwidth is not available. We shall soon see how this arises. In digital transmission, however, we invoke methods to control the distortion of the pulses so that the receiver is still able to distinguish one type from another—between positive and negative pulses in this case. Second, the lack of bandwidth causes some smearing of the pulse shape in time, causing leading and trailing pulses to overlap. This creates *intersymbol interference* whereby the sequential pulses interfere with each other's shape. Again, this is of no consequence if we can arrange the smearing into a form that does not add any ambiguity to distinguishing between the two classes of symbol. Given those considerations, let us look at how our elementary system fares if the bandwidth is strictly limited.

The ideal limited bandwidth channel would have, at baseband, a frequency response flat up to some frequency F and zero from there on. If we were to apply a pulse to this, Fourier transform theory shows that the output voltage waveform at the receiver would take the form given by sin(kt)/t, where k = 2πF. (A short summary of Fourier transform theory appears in Appendix 1). The frequency response and pulse waveform is shown in Figures 2.2.a and 2.2.b respectively. Although this resulting waveform is far from the ideal pulse the transmitter may have sent, it has a clearly defined peak, and the receiver could distinguish positive from negative by sampling at the peak.

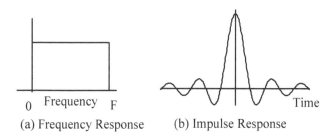

(a) Frequency Response   (b) Impulse Response

Figure 2.2: Channel Frequency and Impulse Response

Now consider two pulses to be transmitted in succession; to be specific a positive pulse followed by a negative one. The manner in which the two output symbols might overlap is shown in Figure 2.3. Notice that if we choose the time between the pulses correctly, the times at which each attains its peak value can be arranged to fall at a time at which the waveform of the other passes through zero. Hence, as the receiver takes samples at these times, the sampled peak value of each symbol is not influenced by the presence of the other. The intersymbol interference is thus zero at these sampling instances.

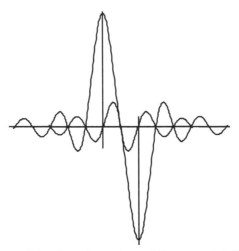

Figure 2.3: Overlapping of Sequential Symbols

From the mathematical form of the symbol waveform given above it is readily shown that the correct interval T between pulses is given by the expression

$$T = 1/(2F)$$

and the rate at which we can transmit symbols over this channel without ambiguity is equal to 2F per second. This is one of the classic results in digital transmission theory, and it arises from simple principles, as we have seen.

So far this argument has centered around transmitting only binary numbers where each symbol needs to take on only two distinguishing levels at its peak, positive and negative. However, the situation is essentially the same if the transmitted pulses are to take on a number of discrete levels. The requirement is still that the peak of each waveform, in time, falls on the zero crossings of all the others. Thus if each pulse could take on one of eight discrete levels, each symbol could convey three bits of data and the data rate would be three times the symbol rate given above. In general, if N levels are allowed, the data rate is $\log_2 N$ times the symbol rate. The task of the receiver is to distinguish which of the N levels was actually sent.

This model, as presented so far, contains a hidden assumption that does not accord with normal practice. This is the condition that the data to be transmitted arrive at the input to the channel in the form of very narrow voltage spikes, or impulses, so that each excites the impulse response of the channel at the output in the receiver. In practice, of course, the data are more likely to be presented as a stepped rectangular waveform with the steps having the full duration T defined above as the time between successive symbols. The major consequence is that, whereas data in the form of impulses would present a flat frequency spectrum at the input to the channel, real data present a falling frequency characteristic in accordance with the well-known $\sin(\omega T)/\omega$ Fourier transform of a rectangular waveform. Practical systems therefore include a pre-equalizer in front of the input filter of the channel to compensate for this frequency characteristic. The situation is illustrated in Figure 2.4. The pre-equalizer ensures that the time response of the overall transmission channel to rectangular input waveforms retains the symbol shape described above, i.e., being equal to the ideal impulse response with zero crossings at the required time intervals. All practical data transmitters will be found to contain pre-equalizers of this nature.

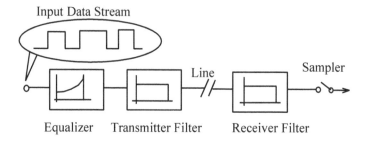

Figure 2.4: Pre-equalization for Rectangular Data Pulses

Spectrum Shaping: The Nyquist Criterion

While the pulse shape in the elementary system described so far is satisfactory to avoid intersymbol interference, the problem remains that the required filter characteristics are idealized "box car" or "brick wall" functions and not realizable in practice. A method is required that will permit the rolling off of the bandpass shape at its edge, and at the same time preserve the zero crossings of the symbol waveforms at the receiver sampling intervals at multiples of time T as is evidently necessary from this discussion.

It can be shown that the desired symbol zero crossings are preserved if the frequency characteristic is rolled off with some function that is antisymmetric around the original cut-off frequency F, as illustrated in Figure 2.5. This is known as the Nyquist criterion, after the famous researcher in the Bell Laboratories who first derived it. Apart from being symmetric about F the shape of the rolloff is fairly arbitrary. It needs to be readily approximated by realizable filters. It should allow the tails of the symbol waveform to decay fairly quickly to reduce the number of symbols contributing to sampling errors when there is a sample timing error in the receiver. In practice, a rolloff approximating a sinusoidal shape is often adopted and referred to by the term "raised cosine" as the resulting form of the frequency response, say H(f), can be written

$$H(f) = 1 \ \text{ for } 0 < f < (1 - \alpha).F$$
$$H(f) = \tfrac{1}{2} + \tfrac{1}{2}\cos[\pi(F - f)/2] \ \text{ for } (1 - \alpha).F < f < (1 + \alpha).F$$
$$H(f) = 0 \ \text{ for } f > (1 + \alpha).F$$

where α is some fraction of the ideal bandwidth F. Large values of α tend to give sharper symbol shapes, easing the matter of synchronizing the sample timing at the receiver, but of course demanding more bandwidth. In broadcasting systems the latter tends to be the more critical, and values of α of only a few percent are found.

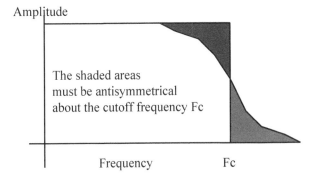

Figure 2.5: Nyquist Shaping of the Transmission Channe

Matched Filtering

The necessary frequency characteristic of a transmission channel carrying baseband digital signals is now evident. In practice, this characteristic is realized by sharp filters at the transmitter and receiver. The filter at the transmitter is necessary to limit the bandwidth of the transmissions, where excessive bandwidth would constitute spurious radiation in radio and broadcasting applications. That at the receiver is required to limit the bandwidth of noise entering the receiver to that strictly necessary, and hence to maximize the signal-to-noise ratio. Both together must constitute the overall desired channel filter response. The question arises, therefore, as to how this filter response should be apportioned between the transmitter and receiver filters.

The result is simply stated but difficult to prove. The filters should be identical, each contributing, in a sense, the square root of the desired overall channel characteristic. This case maximizes the ratio of the midpoint amplitude of the symbol waveform, at the receiver sampling instance, to the level of noise admitted to the receiver. Such filters are said to be *matched*. It follows from this also that the term "root raised cosine" often appears in the description of transmitter and receiver filters, since the desired channel characteristic is frequently of the raised cosine type described above.

Noise and Error Probability

Consider a system intended to transmit symbols with some number N of defined amplitudes and therefore conveying $\log_2 N$ bits per symbol. With the channel characteristics designed as above, and if the channel were noiseless, the midpoint amplitudes of the symbol waveforms must lie at N clearly defined voltage levels at the receiver sampling instances. Noise adds to these levels in the form of a spurious voltage. Accordingly, the sample values in the receiver exhibit a scatter about their true levels. The task of the receiver is to decide, given a sample value, which symbol amplitude was the one most probably sent. The situation is shown diagrammatically in Figure 2.6. Clearly, the most probable value is that which lies closest to the sampled voltage. Thus, the receiver has decision thresholds at well-defined intervals as the diagram shows.

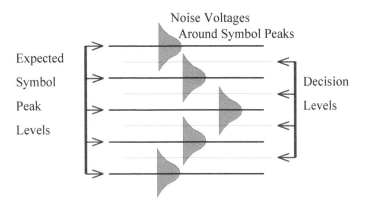

Figure 2.6: Added Noise and Decision Levels

In Figure 2.6, the decision thresholds are shown spaced by a voltage V and the noise voltage is shown as having a statistical distribution with a small standard deviation. The probability that the noise voltage will exceed the value V/2, and hence cause an error, is readily calculated by normal statistical techniques. As V is closely related to the received signal level, and the noise voltage distribution to the noise power admitted by the receiver, the probability of error can be expressed directly as a function of the signal-to-noise ratio. Figure 2.7 shows the typical form of this relationship, in this case for a binary baseband system where the received symbols would have only the two values +/-V.

In this figure the value σ is formally the standard deviation of the noise voltage, as defined by its statistical distribution. σ^2 is a direct measure of the noise power, as V^2 is of signal power. Thus these diagrams are often found with the x-axis related to signal to noise *power* ratios. Now to complicate this a little further, since both the noise power and the symbol rate are both proportional to bandwidth, the energy per symbol and the noise power in a one hertz bandwidth are similarly both proportional to the signal and noise voltages. Thus, one also finds x-axis expressed in terms of the ratio of bit energy to noise density. All these, however, are only different ways of saying the same thing. Generally, and even for multilevel symbols, these diagrams all have the same essential shape.

Figure 2.7: Error Probability versus Noise Level

For broadcasting purposes the assumption of steady noise statistics is a little oversimplified, although it may be useful as a comparative measure of the robustness of competing modulation systems. The character of the noise is more often in the form of impulses or spikes from power lines and ignition systems or in sporadic interference from distant transmitters. Hence errors tend to occur in bursts, rather than randomly, and all broadcasting systems include methods of handling such events, as we shall discuss under the topics of radio propagation and forward error control.

The Principles of Modulation Methods

Over the years there have been many clever methods for impressing digital information onto a RF carrier, including amplitude phase and frequency modulation. Of these, only amplitude modulation remains important in broadcasting (and in a number of other applications as well). It comes in two guises, direct amplitude modulation, which encompasses vestigial sideband transmission (VSB), and quadrature amplitude modulation, or QAM, which has application in both terrestrial and satellite systems.

In the most general terms, a RF carrier can be considered as comprising a cosine component, written as $\cos(2\pi ft)$, and a sine component, $\sin(2\pi ft)$, often referred to as the in-phase and the quadrature component respectively. These can be linearly modulated with separate baseband signals, say I(t) and Q(t), which may or may not be related. Thus, the general modulated signal, say s(t), can be expressed as:

$$s(t) = I(t) \cos(2\pi ft) + Q(t) \sin(2\pi ft)$$

The concept is shown on a voltage phasor diagram in Figure 2.8. Figure 2.8 illustrates the convention by which I(t) and Q(t) are also called the in-phase and quadrature modulation components respectively.

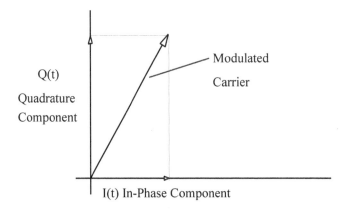

Figure 2.8: In-Phase and Quadrature Modulation

In direct amplitude modulation only the first term is applied, and I(t), for example, would be the stream of symbols from the baseband system described above. The Nyquist spectrum shaping and matched filtering can be applied at baseband or IF. As in analog TV transmission, the occupied RF bandwidth is actually twice what is strictly required, and in the ATSC system using this technique, VSB filtering is applied at the transmitter in a very similar manner to that applying in the analog world.

In QAM systems, I(t) and Q(t) are both streams of baseband symbols like those discussed above, and subject to the same requirements for Nyquist spectrum shaping and matched filtering. However, in this case I(t) and Q(t) are both derived from the one input stream and strictly related so that, at the receiver sampling instances, the modulated carrier can take on only a finite set of values over the in-phase and quadrature domains. These values can be plotted in a diagram where the horizontal axis represents the in-phase component and the vertical axis, the quadrature component, both at the receiver sampling instance. These are known as "constellation" diagrams. Figure 2.9.a shows an example where I(t) and Q(t) can each take on four levels, and the resulting carrier may occupy 16 different positions. Not unnaturally, this is called 16 QAM. Figure 2.9.b shows the diagram for 4 QAM. For obvious reasons this is usually, but inaccurately, referred to as QPSK, standing for quadrature phase shift keying. Constellation diagrams can of course, encompass direct amplitude modulation, and that for 8 levels appears in Figure 2.9.c.

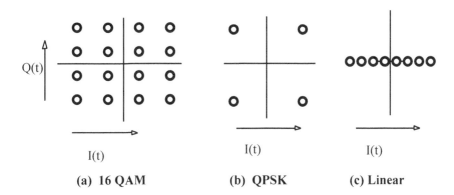

(a) 16 QAM   (b) QPSK   (c) Linear

Figure 2.9: Constellation Diagrams

Constellation diagrams are useful for describing system performance and testing. Thus if noise and distortion are imposed onto the received signal, the constellation positions that the carrier can occupy become spread out and blurred, as shown for QPSK in Figure 2.10. The receiver's task at any sampling instance is to determine which of the constellation positions was the one actually transmitted. It usually does this by choosing the one that would lie closest to the position of that being received. Obviously, if the spreading of the received signals around the allowed positions is such that overlap occurs, there is a significant probability that an error will occur in this decision.

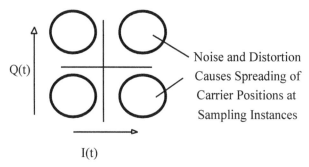

Figure 2.10: Noise and Distortion on Constellation Diagram

Evidently, the noise immunity of a system is directly related to the distance by which the points on the constellation diagram are separated. This is often referred to as the "Euclidean Distance," and will be discussed further in the chapter on error correction.

On the constellation diagram, one could draw a small line from a correct clean constellation position to the position of a symbol actually received. This would constitute a small vector on the diagram, and a statistical summation of these vectors over a large run of symbols gives a measure of how well or badly the system is performing. This is the basis of the "error vector method" of testing, to be described in more detail in Chapter 11.

VSB Modulation

The method of modulation chosen for the United States ATSC digital television standard is vestigial sideband or VSB. At first glance one might think that the preceding discussion involving in-phase and quadrature components is not applicable in this case. In fact, however, viewing VSB in terms of these components can be very useful, particularly in the realms of system analysis and design.

Vestigial sideband (VSB) is in reality a variant of single sideband (SSB) modulation. Many readers with a background in radio techniques will be familiar with the "quadrature method" of generating SSB. Figure 2.11 shows a block diagram of how this is typically realized in practice. Notice that there is an "in-phase" and a "quadrature" modulator. The former imposes a linear amplitude modulation on the in-phase carrier designated by $\cos(2\pi ft)$, whilst the latter modulates the quadrature component $\sin(2\pi ft)$. The signal feeding the in-phase modulator is simply the baseband input, and the signal feeding the quadrature modulator is that same signal passed through a 90-degree phase shifter. A little mathematics shows, relatively easily, that this 90 degree phase shift, together with that implied in the sine and cosine expressions for the carriers, causes the lower sidebands from the two modulators to be in antiphase. Hence, when the modulated signals are added, the lower sidebands cancel, leaving only the upper sidebands. The result is a single sideband signal.

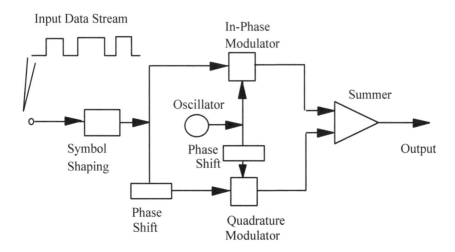

Figure 2.11: SSB or VSB Signal Generation

This scheme is readily modified to produce a VSB signal, particularly now that items like the baseband 90-degree phase shifter are easily realized by digital signal processing (DSB) integrated circuits. The important point being shown here is that the VSB signal does contain in-phase and quadrature components, just as do the various QAM schemes in use. Both components must be transmitted faithfully in VSB for satisfactory system performance. Notice that, however, in contrast to QAM the two components are strictly related and not independent.

In VSB, the quadrature component has a symbol waveform quite different from that of the in-phase sin(kt)/t type in Figure 2.2.b. In fact, the quadrature symbol passes through zero where the in-phase symbol has its peak (i.e. at the sampling instance) and peaks close to where the in-phase symbol has its zeros. Thus the constellation diagram for VSB would ideally appear as in Figure 2.12.a, but in practice transmission impairments always cause some crosstalk, and a more realistic representation would be as in Figure 2.12.b.

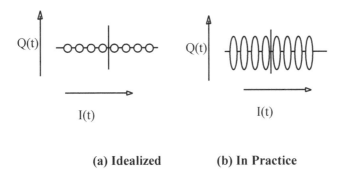

(a) Idealized   (b) In Practice

Figure 2.12: VSB Constellation Diagrams

R.F. Bandwidth and Bit Rates

It is well known that under most forms of linear modulation, but not VSB or SSB, a channel spanning baseband frequencies from zero to F, will occupy an R.F. bandwidth B equal to 2F. This is true whether the modulation involves either the in-phase or quadrature components described earlier. The baseband channel can support a symbol rate of 2F, as demonstrated earlier in this chapter, and thus the symbol rate in an ideal R.F. channel is effectively equal to its bandwidth. As with the baseband channel, however, we must make some allowance for the Nyquist spectrum shaping, so in practice the actual symbol rate is slightly less than the nominal bandwidth.

Generally, but again not for VSB or SSB, the in-phase and quadrature components of each symbol at the receiver sampling instants define its position on the constellation diagram. Thus for QPSK, a symbol can take any of four values and encode two bits, a 16-QAM symbol can take 16 values and encode 4 bits and so on. Hence the bit rates achievable are approximately B, 2B, 4B, and 6B in a channel of bandwidth B according to whether binary, QPSK, 16 QAM, or 64 QAM is chosen. The price to be paid, of course, is the reduction in noise immunity at the higher bit rates.

With VSB (or SSB) the picture is slightly different. A baseband channel of width F translates into an R.F. channel of bandwidth B also equal to F. In principle, therefore, the symbol rate can be equal to twice the R.F. bandwidth. Thus for binary, four, eight, and 16 level modulation the available bit rates under VSB are 2B, 4B, 6B, and 8B respectively.

For a given number of constellation positions VSB offers twice the bit rate of QAM modulation schemes. However, a glance at the constellation diagrams shows that the Euclidean Distances, and thus the noise immunities, are generally greater for QAM. In practice, when a lot of other issues are considered, the ultimate cost and spectrum efficiencies of both systems are much the same. This is reflected in the decision by the United States to adopt one system, and by the Europeans to adopt essentially the other.

Hierarchical Modulation

A number of situations arise in digital transmission where some priority symbols can usefully be given a greater immunity against errors than that which applies for the bit stream as a whole. Frequently, this occurs for the bits that are used to maintain synchronization of the clock controlling the sampling timing at the receiver. If this synchronizing is lost due to a burst of errors, it may take some time for the receiver to reacquire lock, and that can be very annoying to the user.

In broadcasting additionally, the concept of a transport stream, as described in Chapter 4, allows more than one service to be carried on each bit stream. Thus, one may wish to transmit a program with a high bit rate for good quality, and a second version at a lower but more robust bit rate, to cater respectively for good and poor reception conditions in the service area. An example often cited is that a TV program might be transmitted in a high definition format on the lesser protected stream and in standard definition format on the more robust stream. Generally speaking, the less robust stream will have the greater payload bit rate because it uses more levels in the modulation scheme and less overhead is devoted to error correction. Hierarchical modulation permits many variations of this mode of operation where robust and not so robust symbols are usefully included in the same transmitted bit stream.

To illustrate the concept, consider a symbol transmitted by 64 QAM and thus able to represent a group of six bits. Suppose now, we wish to transmit two of those bits, say bits number five and six, with greater security than the other four. Figure 2.13.a shows how this might be arranged on the constellation diagram. In effect, bits one through four determine a position to be occupied in a 16-position constellation diagram, which is a subset of the full 64-position diagram. This 16-position constellation is then placed into one of the four quadrants of the larger diagram. As there are four quadrants, two bits are sufficient to determine which one is to be occupied. In this example, bits five and six are allocated to this function.

(a) Constellation Ratio = 1   **(b) Constellation Ratio = 2**

Figure 2.13: Hierarchical Modulation 64 QAM

Obviously, the receiver has an easier task in deciding to which quadrant the received point belongs, than it has in deciding between two points within a 16-position sub-constellation occupying one of the quadrants. Hence, we see that errors are much less likely to arise in bits five and six than in the four others. These bits can be made even more robust by spacing the smaller sub-constellations even further apart, as illustrated in Figure 2.13.b. The proportion between the normal spacing, as in Figure 2.13.a, and the additional spacing in Figure 2.13.b, constitutes the *constellation ratio*. The standards for transmission systems often permit a choice of several values for this parameter.

Evidently, hierarchical modulation could also be applied to VSB systems, for example by using one bit to determine whether a constellation point falls into the upper or lower half of the diagram. In practice, the ATSC system uses a variation of this, and the modulation becomes binary or two level periodically to define "segment synchs," which synchronize the timing of data packages. This is described in more detail in Chapter 8.

Chapter 3

The Effects of Radio Propagation

As a transmission medium the atmosphere is far from perfect. Three effects are particularly unfriendly to digital transmission. These are signal fading, multipath propagation, and impulse noise. These have, of course, always had to be considered in analog transmission systems. Their impacts in digital, however, are very much worse, in that they are quite capable of causing catastrophic signal failure. Therefore, they need careful consideration in system design. In this chapter, we will review the mechanisms by which these adverse conditions arise and show how they have influenced the design of digital systems. It is probably true to say that a want of measures to combat propagation effects has been one of the controlling factors in bringing digital transmission into the broadcasting field.

Perceptual Signal Degradation

Proponents of digital broadcasting have long argued that the major benefit it will bring to viewers and listeners is the absence of degradation due to poor transmission conditions. Amongst these they include audio and video noise, ghosting, airplane flutter, and interference from power lines. All this is very true, as long as the transmission parameters remain within some envelope carefully chosen by the system designer. Should the parameters fall even slightly outside this envelope, the result is likely to be disastrous for the received service quality. A short explanation of what happens under a condition of falling signal to noise ratio will serve to illustrate this point.

When signal fading occurs in an analog system the net effect is to decrease the received signal to noise ratio and increase the perceived level of video or audio noise. As fades deepen, the picture or audio quality gradually deteriorates, but even in relatively deep fades the picture remains watchable and the audio understandable. Analog systems are therefore said to fail "gracefully." Most people find this tolerable as long as the quality is good most of the time.

On the other hand, digital systems fail catastrophically. Recall that each pixel value or sampled audio level is converted to a binary number and transmitted as a sequence of bytes comprising a binary word. Thus if the error rate is small, most numbers will be recovered correctly at the receiver, with the occasional one being interpreted incorrectly. In TV, the picture will be essentially perfect over the screen with only the odd dot showing from a pixel with incorrect brightness or color. In audio, only an occasional click will be perceived. However, all systems contain forward error correction, and at moderate error rates even these artifacts will not be present and the perceived service quality will remain perfect, even with moderately high error rates. This is why digital transmission is said to avoid the degradation that plagues analog systems under adverse propagation conditions. There are two reasons, however, by which this desirable characteristic suddenly fails.

First, consider how the underlying bit error rate behaves with a falling signal level. From the graph of error rate versus signal level in Figure 2.7, we see that at reasonable error rates any small reduction in signal level will cause a great increase in that error rate. It is not unlikely that a fall of one dB in signal to noise ratio could increase the error rate by an order of magnitude. In systems without error correction, therefore, there is a relationship that gives many dB's increase in video or audio noise level for each dB of signal decrease. In analog systems, that relationship is generally one dB per dB. The digital effect does not matter a great deal, as long as the error rate is low enough to keep the level of audio or video noise below the level that can be perceived by humans. Once it passes that perceptual threshold however, its very rapid increase quickly makes it intrusive to the listener or viewer. In systems with error correction, the onset of intrusive degradation is likely to be even more sudden.

We shall see in Chapter 5 that forward error correction can correct only up to a certain number of errors in each binary word. If the number or errors received is below this number then the reception will be perceived as perfect. For example, say that number is 10. Then, if 11 errors occur, the system is unable to make the necessary correction. By virtue of the manner in which error correction operates, the receiver is likely to interpret the binary word as totally, rather than slightly, different from that transmitted. Hence, the decoded output level will also be totally different from that transmitted. Note that this is much worse than what happens with added noise in analog systems. That at least leaves the output level usually somewhere close to the value transmitted. At the point where the forward error correction system is about to fail, the system will be operating at a significant underlying error rate. Hence, when it does fail the effect is likely to be a dramatic difference between what was transmitted and what is received.

Figure 3.1 summarizes all this by contrasting the manner in which the perceived service quality fails in analog and digital systems. This diagram illustrates why the signal level for digital systems must stay above some critical level for a much greater proportion of locations and times than is necessary for analog. Some viewers may put up with a noisy picture for 10% of the time, but they certainly would not accept having an unusable service to anything like that extent.

Figure 3.1: Contrasting Failure Modes

Signal Fading

Those who live on the outer fringes of a broadcasting service area know that the reception quality can be quite variable. It can vary from location to location, in that where one household might enjoy quite good reception, neighbors a few streets away might have to use elaborate antenna systems to obtain even a just reasonable service. The quality can also vary with time and is particularly influenced by weather conditions.

Location variability within small local areas is generally caused by what is known as *clutter loss*. The name reflects the cause, which is quite literally the blocking of the transmission path by the clutter of buildings, trees, and local hills to be found in the vicinity. This loss tends to be fairly stable but is influenced particularly by rain wetting trees and the surfaces of buildings. That is largely a problem for the antenna installer at the receiving end, and system designers must be concerned with issues on a larger scale.

System designers must take into account the variability of signal level both in location and in time over entire service areas. To assist them both the Federal Communications Commission (FCC) and the International Telecommunications Union (ITU) have published graphs that show the signal levels to be expected in typical situations.

The curves take the form shown in Figure 3.2, which is a very simplified version. They are plots of the expected level of field strength against distance for one kilowatt radiated from various transmitting antenna heights. They are universally known as the F(L,T) curves where F denotes field strength, L is the percentage of locations, and T is the percentage of time for which the field strength will be greater than the value plotted. Thus the F(50,50) curves show the field strengths that will be exceeded for 50% of locations and 50% of the time at various distances in typical terrain. Correction factors are available to adjust the values for differing terrain characteristics, for example for propagation over water or hilly districts. The curves have been derived from theoretical analyses adjusted to take into account many years of measurements in the field.

Figure 3.2: Typical F(L,T) Curves

In analog system design the F(50,50) curves are the ones normally used to determine the necessary transmitted power level. However, in setting the parameters for the United States digital TV plan, the FCC used the F(50,90) curves, and many people argue that digital system design should be based on F(95,95) or even much more severe requirements.

An analysis of these curves can be quite revealing in terms of the variability one might expect in typical service areas. Figure 3.3 shows plots of signal variability for VHF and UHF services against distance from the transmitting site. The plots are obtained by finding the difference in signal levels between the F(50,50) and the F(50,1) curves. Signal levels are usually found to be symmetric about the 50% level, at least in time (i.e., statistically, this assumes that the level has a log-normal distribution.) Hence, Figure 3.3 can be interpreted as showing the difference between the level available for 50% and 99% of the time. The upper curve applies at UHF frequencies and the lower at VHF frequencies.

Error! Not a valid link.

Notice that according to this figure fading is not likely to be a large problem out to about 50 km. It worsens rapidly beyond that for UHF frequencies. On the edge of large service areas, differences as great as 20dB could exist between the levels available for 99% and 50% of the time. The corresponding figure is about 10 dB for VHF services. Analog transmitted power is usually calculated according to the F(50,50) curves, but digital reliability should aim at something close to 99% availability. Thus, the variability illustrated in Figure 3.3 has serious implications for transmitter power, at least in large service areas. Each situation should be examined individually, and to this end a review of the propagation mechanisms involved may be useful.

Atmospheric Refraction

The principal cause of the variability in time is the effect of atmospheric refraction over the transmission path. Consider a typical situation shown in Figure 3.4.a. This shows a ray line from a transmitter to a domestic receiving antenna. The ray line is drawn as a straight line to show the situation when the atmosphere is uniform. Notice that it clears the obstruction drawn at about midpath.

Figure 3.4.a: Clear Ray Line, Normal Refraction

Now let us suppose the refractive index of the atmosphere changes with height above the ground. This happens whenever there is a temperature gradient, as the refractive index of air is temperature dependent. Recall also that the refractive index governs the speed of propagation of electromagnetic waves, light or radio signals, through a medium. Hence, if the variation in refractive index causes, for example, the lower edge of the wave path to have a greater velocity than the upper edge, there will be a tendency for the wave to curve upward. The effect is that a ray line between the transmitting and receiving antennas must take a concave upward path as shown in Figure 3.4.b. Notice now that the ray path could now be partially blocked by a midpath obstruction as shown in the figure.

Figure 3.4.b: Ray Path Curved by Refraction

To take this example one step further, Figure 3.4.c shows Figure 3.4.b redrawn and distorted in such a way that the ray path appears straight. The distortion has the effect of increasing the apparent curvature of the earth, or equivalently, making the radius of the earth seem smaller than it actually is. Hence, the literature on propagation often refers to a "k factor," where k is the number by which the radius of the earth would have to be multiplied in diagrams like Figure 3.4.c to make the ray path appear straight. Notice that the decreasing earth radius on this diagram has the effect of pushing upwards the profile of any potential obstruction, especially close to the midpath position.

Figure 3.4.c: Earth Curvature Shown Increased

This explanation will be seen as a little oversimplified, as it must be obvious that the energy propagating from the transmitting to the receiving antennas cannot be confined to the single line we have called the ray line. In fact, the energy occupies a volume of space contained within what are known as the Fresnel zones, named after a famous researcher into electromagnetic wave propagation. Anything that penetrates into the Fresnel zones causes some loss of received signal level. In broadcasting, the earth's bulge and ground obstructions always penetrate into the Fresnel zones, and the propagation loss is greater than it would be in free space. Hence almost any variation in atmospheric conditions, by causing a change in k, will cause the zones to be more or less obstructed and cause a variation in received signal level. Figure 3.5 shows a graph of the cumulative probability of k in a temperate climate.

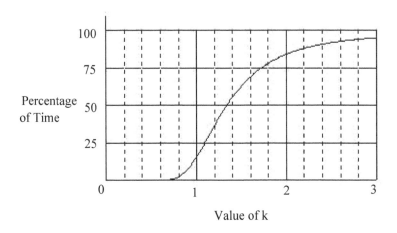

Figure 3.5: Cumulative Probability of k

The extent to which k changes largely governs the degree to which the signal strength varies over any propagation path. Normally k has a value of around 1.4 and varies over a range of approximately 0.8 to 2. Notice that k falls below 0.8 for only a very small percentage of time, although even that percentage may be significant in the context of digital signal reliability. Whether a path becomes obstructed or not depends on its geographical features of course. Table 3.1 shows the effective increase in height of a midpath obstruction as k varies, in the sense of Figure 3.4.c, with k = 1.4 taken as reference.

Value of k	Path Length km		
	75	100	125
1.4	0	0	0

1.2	13 m	23 m	36 m
1.0	31 m	56 m	87 m
0.8	59 m	105 m	165 m

Table 3.1: Increase of Obstruction Height with k

The main effect of a small k is often to make the earth's bulge penetrate the ray line, and this effectively puts the receiver beyond the radio horizon. In this context, Table 3.2 shows how the radio horizon contracts as k becomes smaller.

Value of k	*Tx. Height 300 m*	*Tx. Height 600 m*
1.4	87 km	117 km
1.2	80 km	110 km
1.0	73 km	98 km
0.8	65 km	90 km

Table 3.2: Contraction of the Radio Horizon with k

Super-Refraction or Ducting

So far the effects we have been considering arise from simple variations in the lapse rate with height of the refractive index in the atmosphere. Sometimes, however, when temperature inversions occur, the lapse rate of the refractive index may exhibit a discontinuity that can, as we shall show, open up a second ray line—carrying the diffracted ray—between the transmitter and receiver. Signals on the direct and diffracted paths interfere, often destructively, giving rise to potentially destructive fading. The region of discontinuity is often called a *duct*, since under some circumstances it can guide waves over very long distances. This form of fading variously goes under the names of *anomalous propagation, super-refraction, ducting* or *duct fading.* The essential mechanism by which it occurs is shown in Figure 3.6.

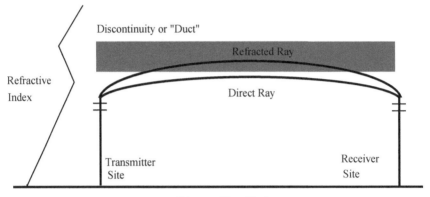

Figure 3.6: Super-Refractive Propagation

Both the direct ray and the refracted ray are drawn curved. The extent of the curvature depends on the lapse rate of the refractive index in the atmosphere. Hence, the upper ray has a greater curvature than the lower in this case, and interference between the two can occur as shown in the diagram. The reason for the term "ducting" is quite apparent. It has not previously been of much concern for broadcasters except on long distance microwave links or transposers. However, with digital working and the strict requirement for very small outage times, it may well become more important in the future.

So how bad is it in terms of causing outages? The author has recently carried out a study utilizing the techniques traditionally applied to microwave systems. This showed that deep fades can occur for small percentages of the time on paths as short as 30 km. The problem is worst at UHF and, of course, on the longer paths. The most important results are contained in Table 3.3. This shows the shortest path length for various frequencies for which 10 or 20 dB fades can occur for more than 0.1% of the time.

0.1% of the Time in the Worst Month		
	10 dB Fade	20 dB Fade
Band IV/V	30 km	60 km
Band III	42 km	90 km
Band I	65 km	130 km

Table 3.3: Limiting Path Lengths for Duct Fading

Multipath Propagation

Multipath describes the situation whereby a signal reaches a receiver by more than one propagation path from the transmitter. Normally, the main signal reaches the receiver by the shortest direct path, and a second signal arrives later, having taken an indirect path by virtue of being reflected off some large object situated away from the main path. The duct propagation just described is a particular example as well. The delayed signals are often described as echoes. They cause ghosting in analog TV and can give rise to severe distortion in FM services, as anyone can attest who has driven through the centers of large cities where many reflections from buildings are often present.

Digital systems are very sensitive to multipath. In Chapter 2 we showed the form that the transmitted symbol waveform must take to ensure reliable detection at the receiver. In particular, the waveform must have a large central peak surrounded by zero crossings at specified instances in time before and after that peak. The signaling rate is arranged so that the zero crossings of leading and trailing symbols fall at the midpoint in time of the central peak of any given symbol. Thus, leading and trailing symbols do not interfere with this central peak, and its level can be accurately determined at the receiver.

Now consider what happens when an echo is present. The peak of a symbol in the echo may fall effectively anywhere, including close to the central peak of some symbol sent earlier. It may thus cause the detected level of the wanted symbol to change considerably, possibly leading to an error and certainly leading to errors being caused by a lower level of noise than would otherwise be the case.

To be specific, let us consider a system in which the symbol amplitudes may take eight discrete values as in the ATSC system used in the United States. The normalized symbol amplitudes will be +/-7, +/-5, +/-3, and +/-1 V and the receiver decision levels will be spaced ½ V from each of these. Now suppose we transmit a +7 V symbol followed by a +1 V symbol, but because of multipath, a delayed and attenuated echo of the 7 V symbol comes to overlay the 1 V symbol. This will definitely cause an error if the echo magnitude exceeds ½ V. Hence the echo attenuation must be better than 14 times or 23 dB. More particularly if the echo is ¼ V it will reduce the signal-to-noise ratio by 6 dB. This is an echo attenuation of 30 dB, comparable to the level of ghosting tolerable in analog TV. Given the failure characteristics of digital systems described, it represents an intolerable loss in service area or reliability.

Note that the time between symbols in the ATSC system is 93 nanoseconds, and typical echo delays in urban environments are some tens of microseconds. Hence, the situation described above can readily arise, and all digital terrestrial broadcasting systems require some method of combating multipath. This is not so true for satellite systems as the receiving antenna beamwidths are usually sharp enough to reject reflected signals.

The two major modulation standards defined for terrestrial transmission are the ATSC eight level vestigial sideband (8VSB) system and coded orthogonal frequency division multiplexing (COFDM). The methods of combating multipath in these will be dealt with in chapters 8 and 9. Briefly, however, the ATSC system uses an active equalizer that inserts echoes at the receiver in the inverse sense to those coming in on the transmission path and thus brings about their cancellation. The transmitted signal contains a defined "training signal" for the equalizer. By subtracting a local version of this signal from that received, the equalizer can determine the character of the echo signals present. COFDM uses very long symbols and adopts a "guard band" in time, during which it simply ignores the incoming signal. The guard band is long enough for the echoes of leading and trailing symbols to have died out for typical urban multipath delays. Hence we see that combating multipath is a very significant feature of practical systems. One can make the definitive statement that without it no digital terrestrial system would work.

Impulse Noise

Treatments on digital transmission in communications systems usually focus on its immunity to steady electrical noise, called thermal noise after the mechanism causing it, or Gaussian after the famous mathematician who gave his name to the statistical distribution characterizing it. This noise has the advantage of being well understood and fairly easily analyzed mathematically. In broadcasting, however, we have to be concerned with the more insidious impulse noise that pervades most modern towns and cities.

Impulse noise has many causes, including motor vehicle ignition, electric motor starters, and many industrial processes. A particularly important source is surface leakage across power line insulators, where the noise comes in bursts at the power line frequency. It is often seen on analog TV sets as a band of noise drifting slowly up or down the screen. Its level is largely unpredictable, as is the probability that it will occur in any particular location. Impulse noise is an annoying nuisance in analog transmission, but rarely causes reception to be unusable. It will, however, often cause intolerable interruptions to digital transmissions.

Chapter 4 on transport streams shows that a significant portion of the digital transmission stream is given over to synchronizing signals of some form. All digital receivers have to establish synchronization at a number of levels. All this takes time and, for example, when a digital receiver is first switched on, it may take several hundred milliseconds to establish lock and produce the desired program output. Relocking is also generally required when changing channels. One of the major challenges in designing domestic receivers has been to make this relocking sufficiently fast to avoid annoying viewers

The transmission of the synchronizing signals is usually made more robust than that of the service content. The aim is to ensure that if a transmission degradation does occur, the synchronization will be preserved even if the picture and audio fail. Relocking should thus be required as infrequently as possible. However, given the unpredictability of impulse noise there is no guarantee that it will still not disrupt the synchronizing signals from time to time.

Impulse noise characteristically comes in short bursts, and when it occurs, it is likely to take out a lengthy string of received symbols. One way of countering its effects then, is to use forward error correction that is effective over a run of several errors. Most forward error correction, however, is effective only for single errors or relatively short runs of errors. Hence, the second technique is to shuffle the order in which the payload bits are transmitted and to reverse this shuffling at the receiver. The shuffling is made as random as possible. Then if impulse noise strikes, the errors will be contiguous in the received stream, but after the bit stream is reverse-shuffled, the errors become randomly scattered through the data stream. Their spacing depends on the number of bits involved in each shuffled group, and may be quite large. The forward error correction circuits in the receiver then readily correct what appear to be randomly distributed single errors.

This technique greatly improves the robustness of the transmissions to impulse noise. It is defined for all current terrestrial system standards.

Chapter 4

Transport Streams

This chapter shows that even the most elementary digital transmission systems require the transmitted information to be organized and identified in some way. When the system is carrying more complex systems, such as video, several audio channels, and perhaps data channels as well, it becomes imperative that there is some way of associating the bits with the correct service. This is the process of multiplexing and demultiplexing, by means of which bit streams for a number of services are combined for transmission over a single link and then separated for delivery to the appropriate output device.

In program distribution, before they reach the transmitter, the broadcasting services may pass through a variety of transmission networks, including optical fiber, microwave systems, and satellites in which they may be multiplexed with and demultiplexed from other digital streams any number of times. Hence, the chosen method of identifying program segments in the broadcasting stream must be compatible with national and international standards for multiplexing broadband services. The two concepts we will introduce here are those of *packetized elementary streams* and of *transport streams*.

The Need for Packets

The most elementary digital transmission system that one might conceive of is that for transmitting sampled values of a mono audio service over a dedicated channel. For simplicity, let us suppose that each sample must be transmitted as four binary bits and hence is approximated by one of the 15 values +/-7, +/-6, . . , 0. If some sequence of values were 6, 3, 2, 1, 5, 7, 2, 5, 3, say, the transmitted bit stream would be 0110, 0011, 0010, 0001, 0101, 0111, 0010, 00101, 0011. Now, let us take out those commas, and the bit stream becomes 011000110010000101010111100100-01010011. What is the receiver to make of that? How can it tell where one number ends and the next one starts? Obviously, we must put back those commas. However we don't need to put back all of them. We could, for example, send the numbers of groups of three in which case the bit stream would become 011000110010, 000101010111, 0010001010011. The comma still turns up regularly and by finding it in the bit stream the receiver can remain synchronized to the transmitted bit stream and determine the starting position of each transmitted number.

At this stage we need some way of transmitting the commas, which obviously must be an easily detected string of bits. In this particular example the sequence 1111 may suffice, as being at the extreme range of our audio sample values, it turns up rarely in the natural range of numbers to be transmitted. However, if it is inserted after every third number like the comma above, the receiver can be designed to seek it out at regular times.

In the terminology normally used for digital transmission, each group of four digits would be referred to as a byte, and the sequence 1111 as the synch byte, where synch is short for "synchronizing." To be strictly accurate, the term *byte* is normally applied to strings of eight bits, but we use four in this example to keep the discussion easier.

The normal method of synchronizing the receiver timing to the synch bytes is by means of circuitry known as *phase locked loops*. The principle is shown in Figure 4.1.

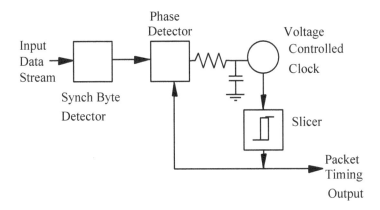

Figure 4.1: Synch Byte Timing Recovery

Broadly, a synch byte detector produces a pulse whenever the synch byte is detected. Its output signal consists of a string of pulses spaced at the regular periods at which the synch pulses are inserted at the transmitter, but with the odd random pulse occurring when a data byte happens to match the synch byte. At the same time a voltage-controlled clock produces a regular train of pulses whose period closely approximates the expected period of the synch bytes. As the name implies the voltage-controlled clock is an oscillator whose frequency can be changed by applying a control voltage. The two trains of pulses enter a phase comparator for which the output voltage is proportional to the phase, or timing difference, between them. It is easy to design a detector that is sensitive to the regular synch byte pulses and ignores the irregular false pulses due to data bytes. The voltage out of the phase detector is fed back to control the frequency of the voltage-controlled clock. This serves to pull the frequency so as to correct the detected phase error between the clock pulses and the synch byte pulses, so that the clock is said to be locked to the synch bytes. This serves to define the master clock timing in the receiver. The timing for all byte detection and interpretation is taken from this clock.

Packetized Elementary Streams

We have just seen that even the most elementary data stream needs to be organized into packets separated by easily recognized synch bytes. However, once the packets are recognized, there is no need for all the bytes in them to actually be numbers representing sample values (audio samples in this example). In practice the number of bytes in a package would be vastly greater than the three used in the example and a few of them could be dedicated to conveying information to the receiver.

To persist with this elementary example, suppose the basic audio transmission system is to be extended to a stereo mode. One possibility is to make the packages carry alternately samples from the left and right channels. We need to tell the receiver which channel each package belongs to, and this might be done by making the first of the three bytes in each package either of two special bytes representing left or right. In the receiver, simple circuitry looks at the first byte and switches the output to the left or right channel as appropriate. Importantly, the receiver is designed to use this control byte for some specific purpose and not as a sample value. In reality, there may be a number of these control bytes in each package. Some current practices are described in the chapters on video and audio compression.

Packaged streams like this are found typically in the video and audio streams in a broadcasting multiplex. Some bytes in each packet may be given over to carrying program associated data, often abbreviated to PAD, which conveys data like song titles or the names of artists. An important component in each package is commonly a time stamp, which shows where that package is related in time to all the others in the multiplex. The multiplex may also support a data stream that may be used to carry almost any information. Each of the streams, which thus make up the multiplex, is known as a packetized elementary stream or PES for short. Each PES typically carries a single compressed video channel, a multichannel audio service, or data in some format.

Transport Streams

In general, each of the PESs that come together to form a multiplex may contain packets of different lengths, as befits the applications concerned (e.g., video, audio, and data in broadcasting). It is desirable, of course, that the method of multiplexing all these services together results in a bit stream that is itself packetized into fixed-length packages with its own unique synch byte like that described in our elementary example.

The way this multiplexing is carried out is illustrated in Figure 4.2. The elementary streams enter the multiplexer in parallel. The stream leaving the multiplexer consists of a sequence of packets of some fixed number of bytes.

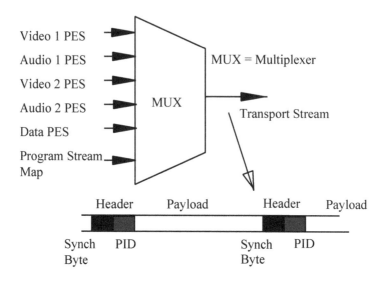

Figure 4.2: Multiplexing of Elementary Streams into the Transport Stream

Each packet has a synch byte and a short header containing information for the receiver. The header is followed by the "payload bytes," a sequence of bytes taken from one of the elementary streams. Part of the header is a program identifier, or PID, which identifies the PES from which the payload bytes have come. One of the PESs, called an elementary stream map, carries data indicating the relationships between the other elementary streams and how they may go together to make up programs. For the situation in Figure 4.2, for example, the elementary stream map would indicate that video 1 and audio 1 make up one TV program.

The lengths of the packets in the individual PESs need not be directly related to the length of the packets in the transport stream. If, say, the PES packet contains enough bytes to fill a whole number plus some fraction of transport stream packets, that part of the final transport stream package left vacant is filled with dummy data. This process is often known as *bit stuffing*. It introduces some inefficiency in that more than the strictly necessary number of bytes has to be transmitted. However, this is nearly always outweighed by the convenience of having fixed-length transport stream packets and of being able to deal with PES packets of any length.

In the ATSC and MPEG realizations of transport streams each packet contains 188 eight-bit bytes. The header includes a 13-bit PID that identifies the PES being carried and also contains a program clock reference. This latter is used to compensate for time variations that may occur in transmission, as telecommunications carriers change transmission routes from time to time to optimize their traffic loading.

Apart from simplifying synchronization at the receiver as we have seen above, the use of fixed-length packages in the transport stream brings a number of advantages. Receiver designs can be standardized over terrestrial, cable, and satellite systems. Error correction is more easily applied to fixed-length packages. They are easily inserted into higher level multiplexes used by national and international communications carriers. The number of PESs that can be inserted into a transport stream is variable and can be adjusted to suit the bandwidth available. The content of each PES is immaterial, and there is much commonality between TV, audio, data, and even communications applications. In broadcasting, each transport stream could carry a variable number of programs varying over high definition TV to VCR quality depending on the bit rate assigned to each.

The level of packaging just described is appropriate to transmitting broadcast services to the general public. The generation of higher-level multiplexes is, however, very much an extension of the process used to form the initial transport stream. Naturally the identification of the contents of the higher order packages becomes more and more complex, and may extend to carrying addresses as each package may have a different destination. In the end, however, it all comes back to putting the commas into a data stream to identify which byte is which.

Chapter 5

Forward Error Correction

Most digital transmission systems incorporate some form of error detection and correction. It is often a vital component of the system. For example, imagine how unacceptable it would be if downloaded computer programs always contained some undetected errors. In broadcasting, it is not quite that critical, but even the occasional error could have bigger than expected consequences. In Chapter 7 on Moving Picture Expert Group (MPEG) picture coding, it can be seen that one error might propagate over several picture frames, for example. Error control also allows the system to operate at lower signal strengths than might otherwise be the case, and hence contributes to minimizing the transmitter power required for a given quality of service. It also makes a less obvious contribution to the realizability of broadcasting services. For example, it permits the clipping of rare but very high peak voltages in transmitter power amplifiers by correcting the odd error introduced by this event.

This chapter explains the principles behind the most common types of error control used in current broadcasting transmission systems. It introduces block codes and convolutional or trellis codes. Broadcasting standards often use the terms *inner codes* and *outer codes*. The words really refer to where they are found in the transmission system. Thus, the inner codes operate just before the signal leaves the transmitter and just after the receiver demodulator. Outer codes tend to be applied close to the extreme input and output ends of the transmission chain. Inner codes are usually convolutional in nature, with an optimum performance under conditions of steady noise interference. Block codes are usually more effective for bursts of errors and are used for outer codes.

It will be useful in this chapter to use the concept of a binary word. A binary word is merely a string of binary symbols, or bits, that can be any defined length. In practical error control systems the binary words may be hundreds of bits long, but they will be much shorter for the examples that appear in this chapter.

Parity Testing

The most common form of error detection is the use of a parity bit. *Parity* means adding an extra one or zero to the end of a binary word so that the total number of ones contained in the new word is either *odd* or *even*. If odd, we are said to have elected to use odd parity. It is even parity if we elect to make the total number of ones even. To take an example that used to be common in computers, suppose we adopt odd parity by adding a ninth bit to each eight-bit byte so that the total number of ones is odd. Thus we would add a one to the byte 01010000 to make the nine-bit word 010100001. Now if one error were to occur in that word a test on the parity would fail, inasmuch as the total number of ones would become even.

Notice that the parity test cannot identify the bit position at which the error occurred. Hence, it can detect the error, but cannot correct it. Systems using simple parity tests usually employ a strategy called ARQ, or automatic repeat request, whereby once errors are detected the receiver sends a request to the transmitter to repeat the data concerned. This is not possible in broadcasting, of course, and error control must take the form of FEC, or forward error correction, using codes capable of both detecting and correcting errors.

Block Codes

The simplest type of error correction employs an extension of the parity concept. For simplicity, we will look at how this would be applied to very short binary words, although in practice, of course, the words would be much longer.

Suppose we were to break up the data to be transmitted into four-bit words and arrange them in a four-by-four matrix (or block) as shown in Figure 5.1. Call these the message bits, designated as "m" in that figure. Now at the end of each row and at the bottom of each column, add a parity bit so that the word in each row and column has even (or odd) parity. These are designated "c" in the figure, as they are commonly referred to as "check bits" in the literature. We now have a 25-bit block containing both message and check bits. This can be transmitted by some method as a 25-bit word.

$$\begin{vmatrix} m & m & m & m & c \\ m & m & m & m & c \\ m & m & m & m & c \\ m & m & m & m & c \\ c & c & c & c & c \end{vmatrix}$$

Figure 5.1: Block of Message and Check Bits

Imagine that at the receiver this block is reassembled into the same form it had in Figure 5.1. If an error has occurred the parity check will fail on both the row and the column that contain the incorrect bit. Thus, this scheme not only detects an error, but also indicates where it lies. It is simply then necessary to invert the bit in that position to correct the error.

To indicate where the error lies, the decoder has to generate some output word. In its simplest form it might well be a 16-bit word with zeros in all positions except that in which the error occurs, and with that position containing a one. Output words of this nature are known as *syndromes*, derived from the medical usage of the word where the syndrome is the collection of symptoms for a disease.

Notice that if two errors occurred, two rows and two columns would fail the parity check, giving four possible locations for the two errors. Hence, this particular method of coding can correct only one error in each block. In fact, almost the only virtue of this method is its simplicity. There are, as we shall see, much more complex methods of generating the check bits to obtain a better result.

The transmission of the 25-bit word can take almost any form desired. A systematic method would be to transmit the 16 message bits followed by the nine check bits. This overall technique would then, not surprisingly, qualify for the name of a block systematic code.

Block systematic codes usually go under the designation Name(N,M), where N is the total number of bits in a word and M the number of message bits. Thus this example would be designated Name(25,16). The name is often derived from the initials of the inventor of the particular code.

Hamming Distance

We have seen that the simple coding scheme above can correct only one error in each word. Practical codes, of course, can correct many errors. One method of appreciating how the more complex are structured is through the concept of the Hamming distance, named after a very famous pioneer of coding theory.

The Hamming distance between two words is the number of positions in which the two contain different digits. For example, the Hamming distance between the two five-bit words 01101 and 11001 is two, since they differ in the first and third digits.

Notice in the simple code above that with 25 bits it is possible to form a total of 2^{25} or nearly 33.6 million different words. Of these, only 2^{16} or about 65000 will ever be transmitted, since only this many different message sequences are possible with 16 message bits. This means that 99.8% of the possible 25-bit words are never transmitted, and if one should be detected out of the receiver it will contain one or more errors. However, if its Hamming distance from one of the permitted words is smaller than that from some other permitted word, then more probably the former was actually transmitted.

This last statement relies on the fact that one error in a word is much more common than two, two errors much more common than three, and so on. This method of choosing the word most likely to have been sent is an example of maximum likelihood decoding, of which we will meet another example later. It is a very common strategy in coding schemes.

Suppose now that the minimum Hamming distance between any of the permitted words is three. If a word is transmitted and a single-bit error occurs, the erroneous received word would lie at a Hamming distance of one from that transmitted. However, it must then lie at a distance of two from some other permitted word. In this case the receiver makes the correct decision by choosing the permitted word with the smallest Hamming distance from that received. However, if two errors had occurred, it would make the wrong decision, because the erroneous word now lies closer in distance to another permitted word. Thus we see that with code words spaced at Hamming distances of three it is possible to correct a one-bit error in any word.

It should now be apparent that to correct two errors per word, the permitted code words should be spaced at Hamming distances of five, or for three errors, at distances of seven and so on. The clever part of designing a code is to pick out that set of words which all lie at the appropriate Hamming distances and employ only those for the permitted code words.

At face value this would seem to be simple, but in practice it turns out to be a far from trivial problem. To illustrate, it can be shown that in our population of 33.5 million 25-bit code words, the selection of those with Hamming distances of five comprises only 409 members. Hence, this is the maximum number of different messages we could transmit with 25-bit words and still expect to correct two errors per word. That is not very efficient. In this argument lies the reason why practical codes comprise very long code words. Only when there is a large number of bits in each word does a significant proportion of the population lie at Hamming distances of the order of five to ten.

Cyclic Block Codes

Practical decoding schemes at the receiver are, in principle, not very different from that implied by the matrixing process in Figure 5.1. In effect the receiver has to form a matrix from the received word and calculate various relationships over the message and check bits. Alternatively, it might contain a look-up table of permitted code words to determine that with the closest Hamming distance to the one received. We have seen, however, that for the correction of multiple errors, the code words must be very long. Hence, either the computational load or the memory requirements in the receiver can become enormous.

A way around this difficulty has been found by using a group of codes known as cyclic block codes. In effect the only permitted words are those which can be generated by shift registers with feedback from output to input. These registers produce very long trains of bits that repeat after some interval (hence the term *cyclic*). It turns out that with this constraint all the necessary operations like calculating the check bits and forming the syndrome in the receiver can be performed by feedback shift registers, which are both fast and simple to implement. Almost all modern block codes are of this type.

The essential defining idea behind operating on cyclic codes is to cast sequences of digits as algebraic polynomials. For example a sequence with four bits would be represented as $a_3X^3+a_2X^2+a_1X^1+a_0X^0$. The exponent at each power of X reflects its position within the block of digits, and the a_n take the values 1 or 0 corresponding to the value of the bit at that position. Hence, the sequence 1101 would be rendered as $X^3+X^2+X^0$. Once this is done, the blocks of digits can be handled by a type of algebra. The rules of this algebra are very similar to those of ordinary algebra, and it turns out that simple binary arithmetic circuits and shift registers can implement most of the algebraic operations.

Figure 5.2 shows a shift register configured as a divider, and the circuitry is seen to be quite simple. The manner in which the polynomial approach is applied is so ingenious that it is really worth indulging in a little straightforward mathematics to appreciate its elegance.

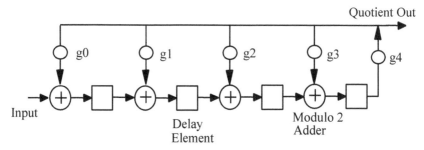

This Circuit Divides the Input Polynomial
by the Polynomial with co-efficients g4, g3, g2, g1, and g0

Figure 5.2: Shift Register Based Divider

Suppose we have a block of bits comprising the message sequence. In the polynomial notation we can express this message sequence as M(X). Now, divide this by another polynomial, say G(X) to give a quotient Q(X) and a remainder R(X). (In Figure 5.2 the shift register contains R(X) at the end of the division.) Send the message M(X) immediately followed by R(X). Let us call this transmitted sequence T(X).

Sending M(X) followed by R(X) effectively shifts M(X) up by n bits – the number of bits in R(X). This effectively multiplies each term in M(X) by X^n.

Therefore,
$$T(X)=X^n*M(X)+R(X)$$
but, with the notation just described, we can write
$$M(X)=G(X)*Q(X)+R(X)$$
whence
$$T(X)=X^n*[G(X)*Q(X)+R(X)]+R(X)$$
and therefore
$$T(X)= X^n*[G(X)*Q(X)]+[1+X^n]*R(X)$$
If the divisor polynomial G(X) is a factor of $[1+X^n]$—i.e., it divides $[1+X^n]$ without remainder—then we see that it will also divide into T(X) without remainder.

At the receiver, the received sequence is divided by G(X). If there are no errors, that received sequence is T(X), and there is no remainder from the division. Any remainder indicates that errors are present. The remainder forms the syndrome in this case, and as with block codes, the error pattern can be calculated from it.

Strictly, the divisor G(X) need only be a factor of $[1+X^n]$, but specific polynomials have been found to have other desirable properties, to the extent that many have been defined in international standards. The reader may see references to these polynomials in system specifications.

The most common of the cyclic block codes are the BCH codes where BCH stands for the names of their inventors, Bose, Chaudhuri, and Hocquenghem. The most famous are the subsets developed by Reed and Solomon. These can operate on a few bits at a time, rather than using the one-by-one bit polynomial algebra we have just considered. Thus in the ATSC digital TV standard we find the RS(207,187) code. Other variations of RS codes turn up in European Standards for TV and radio broadcasting.

Convolutional Codes

As seen, block codes take a group of bits, operate on them to form the code word, dispatch that down the transmission channel, and then take the next group of bits for the same treatment. This might be described in industrial parlance as a batch process. In contrast, convolutional coding is more akin to a continual flow process. Convolutional coders also operate on a group of bits, but the group flows continually through the coder and changes only one bit at a time as the data bits arrive. Thus the effect of any one bit ripples through the output stream for as long as that bit remains in the coder. This ripple-through effect is similar to a mathematical process known as *convolution*, literally a rolling together of the effects of many causes. As we shall see, diagrams that illustrate the process look similar in form to garden trellises, and consequently this coding technique is very often referred to as trellis coding.

Convolutional coders do not output words, as do block coders, but rather a continual stream of bits. However, the bit stream cannot take on all possible sequences of bits. Only certain bit sequences are allowed and those transmitted are determined by the sequence of bits in the incoming data stream. It follows that, since the incoming data stream can encompass all possible sequences, but the outer stream only a subset of that, then the output bit rate must be greater than the input rate. The relationship between these bit streams defines the "rate" of the coding process. Thus a rate 2/3 implies that the message bit rate is two thirds of the coded bit rate and so on.

For the purposes of illustration we will adopt the simple rate one-half coder shown in Figure 5.3. Note that each of the outputs, labeled X and Y, depends on both the incoming message bit and also on the contents of the shift register. The contents of the shift register is conventionally referred to as the "state" of the coder, and one sometimes sees references to these devices as "state machines." In this particular case the coder can be in any one of the four states 00, 01, 10, or 11. For each input bit, the output switch arranges for first X and then Y to be output. Thus, two bits are output for each input bit, giving the coder its rate one-half property. Figure 5.4 is the trellis diagram for the coder.

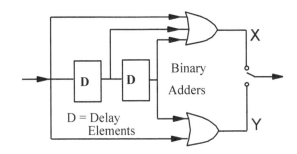

Figure 5.3: Rate One-Half Coder

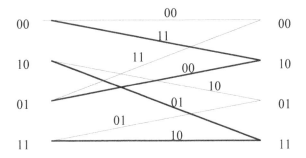

Figure 5.4: Trellis Diagram for Rate One-Half Coder

On the left-hand side are the four possible states of the coder. It will be in any one of these when an input bit arrives. The state of the coder will then change to a new state, shown on the right-hand side of the diagram, with that state depending on whether the input bit is a one or a zero. The possible paths from the left hand to the right hand states are shown by the lines drawn between them, full lines representing a one input and dotted lines a zero input. As stated, the output depends on both the starting state and the input. It is shown by the two-digit byte marked above each of the lines in the figure.

Let us now trace through the sequence of events that occur in the transmitter and the receiver during the transmission of a short string of data bits. To make the explanation simple and the diagrams legible we will confine this to a string of only a few bits and choose the sequence 1010. The situation at the transmitter is shown in Figure 5.5.

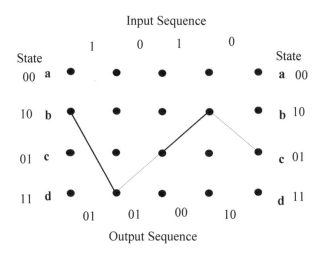

Figure 5.5: Trellis Path in Transmitter

This figure indicates the path through the state trellis for the transmitter starting in state 10 and shows that the output sequence would be 01 01 00 10. In this diagram, the states are labeled **a**, **b**, **c**, and **d** for clarity in the following discussion.

Now let us turn our attention to the receiver. If the receiver is in state 10, the paths through the trellis diagram for all possible sequence are shown in Figure 5.6.a. This diagram is confined to only three shifts as otherwise it becomes unreadable.

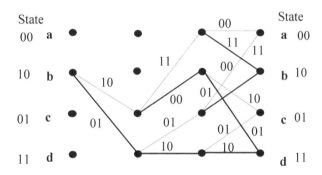

Figure 5.6.a: Possible Paths From State 10

The starting state in the receiver may not be the same as in the transmitter, particularly if there have been errors in the bit stream previously received. Hence, if the receiver were to be in state 00, the possible paths become those shown in Figure 5.6.b. We could draw similar diagrams for the other starting states.

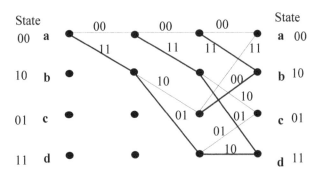

Figure 5.6.b: Possible Paths From State 00

The action at the receiver is to take all possible paths through the trellis diagram and calculate the Hamming distance of each of them from that actually received. For example, suppose some path would generate an output sequence 01 11 00 00. Compared to the output sequence in our example, namely 01 01 00 10, this gives a Hamming distance of two. Of course, if there were no errors, some path must give a Hamming distance of zero. The receiver selects the path with the lowest Hamming distance as being the one most likely to have occurred at the transmitter. Here again is an example of maximum likelihood decoding.

A moment's reflection will show that this is no trivial task for the receiver. In principle, it could operate on sequences of any length, as there are no defined blocks of data in this continuous flow process, and at any time it cannot be sure that it is in the same state as the transmitter. The solution lies in adopting a strategy known as the *Viterbi algorithm*, named after its inventor and one of the most famous principles in coding theory.

The Viterbi Algorithm

The key concept in the Viterbi algorithm is that of "survivors." To explain this, let us take the first six bits, namely 01 11 10, generated by our test sequence. Let us compare this with the outputs that would be generated by a few of the paths in Figure 5.6. The Hamming distance between the output of a particular path and the received sequence is usually called the metric of the path and we shall adopt that in this discussion. Hence, we can now write down the state sequences for a few paths, their output bit streams, and their metrics. Referring to Figure 5.6 we have:

Path **bcab** Sequence 10 11 11 Metric 5
Path **abdc** Sequence 11 01 01 Metric 2
Path **bcbc** Sequence 10 00 01 Metric 4
Path **bddc** Sequence 01 10 01 Metric 3

Notice that the last two paths start at the same state, **b**, and finish at the same state, **c**. Of the two, however, the last has the smaller metric and is likely to be the more correct if there have been transmission errors. The Viterbi strategy is to discard the path with the larger metric. The remaining path becomes the survivor path linking two given states. Thus over this six-bit sequence the survivor path linking the states **b** and **c** is **bddc**. There are, in total, only 16 survivor paths, each linking one of the four starting states to one of the four ending states.

If we were to step Figure 5.6 out one more stage, the then surviving paths must include, as part of themselves, the survivors already found. This argument becomes obvious if one thinks of the trellis nodes as cities and the path metrics as measuring the distances between them. Thus, if we know the shortest route from San Francisco to London and wished to make the journey San

Francisco, London, Paris, or San Francisco, London, Rome we would obviously use this shortest route from San Francisco to London in both cases. (Because of this analogy, the Viterbi Algorithm is often referred to as the shortest path algorithm.) Thus for a real and lengthy sequence, the number of surviving paths never has to be greater than 16. This dramatically reduces the number of possible paths that the receiver has to store and compare with the input sequence.

In practice there is another saving. We said above that the receiver "does not know" the state in which the transmitter started, and so in principle has to retain the details of all 16 possible paths for an indefinite period. However, after a reasonably short sequence has been examined one of the 16 paths will emerge as having a much smaller metric than any of the 15 others and is selected by the receiver as the most probable. Thus the state of the receiver essentially locks in with that of the transmitter after a short time. The necessary number of lattice transitions in a sequence to achieve this is about five times the number of states in the lattice, i.e., about 20 transitions or a sequence of 40 transmitted bits in our example. Hence, the receiver memory need have only enough capacity to accommodate all surviving paths over a bit sequence of this length.

There are variations on the Viterbi algorithm. For example, in one strategy the receiver follows a path while its metric remains small. If the metric increases beyond some preset limit, the receiver backtracks over a few steps and tries a branch different from that previously traversed. These algorithms are comparatively complex and are usually implemented in software on dedicated LSI chips.

Error Probability in Convolutional Codes

Block codes operate by permitting only certain code words to be transmitted, amongst all those possible within the length of code word. The Hamming distance of the received word from those permitted determines which word was most probably sent. The manner in which word errors may occur is easily visualized, and the performance is reasonably easy to calculate.

Convolutional codes operate similarly by permitting only certain sequences of bits, and not all possible sequences, to be transmitted. The permitted sequences are those traced out by legitimate paths through the lattice. If an erroneous sequence is received, the one most probably transmitted is taken as that, of all those possible, which has the smallest metric measured against the received sequence. The essential difference from block codes is that the permitted sequences do not have a defined length and do not form a convenient fixed set of words. Thus we cannot speak of a word error probability and need some other means of assessing performance.

Let us look at what would happen at the receiver in the event of a transmission error. To do this we will consider the transmitter sending a long sequence of bits, and for simplicity will take this as an all zeros sequence. This may not seem like a very typical sequence, but as the argument develops, the reader will see that any sequence could have been chosen and the same result achieved merely by distorting the lattice diagram.

For this sequence, the receiver should stay in state **a**, i.e., the path through the lattice in Figure 5.7 should be **aaaaa--**. However, if an error occurs, the receiver could deviate from this path and take another. Eventually, however, it must come back to state **a**, because if it did not, the metric of the erroneous path would increase without limit. Figure 5.7 shows the shortest two of many possible paths the receiver could take in departing from and returning to state **a**. This illustrates the remark made previously that convolutional coding permits the use of only a subset of all possible paths—those permitted here being only those shown in Figure 5.4

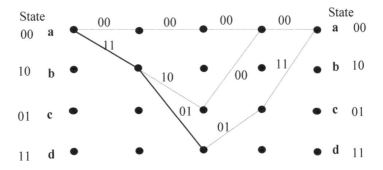

Figure 5.7: Possible Paths Under an Error

On the erroneous path **abca** the imputed bit sequence is 11 10 00, as against the correct all zero sequence 00 00 00. Notice that the Hamming Distance between the paths is three, and we conclude that at least two errors must occur in the bit stream to make the receiver rate the erroneous path more probable than the correct path. For the path **abdca**, the effective Hamming Distance would be five, and three errors must occur to make this the more probable path. Intuitively, one can see that this is a very effective error correction mechanism, as at low error rates, it is highly unlikely that multiple errors will occur in such short bit sequences.

In the literature, the Hamming Distances here are often referred to as the "weight" of particular sequences. A coding scheme is said to have a minimum weight, which applies to the shortest erroneous path permitted and is a good indicator of performance.

Trellis Coded Modulation

A very simple example serves to show that quite some advantage can arise by considering together the characteristics of the forward error correction scheme and those of the modulation system. Take an elementary, but realistic, transmission system that employs four-phase PSK, or 4-QAM, modulation and simple forward error correction capable of correcting one but not two errors in a block. The two binary digits applying to permitted phase of the carrier might be chosen as shown on the constellation diagrams in Figure 5.8.a or 5.8.b.

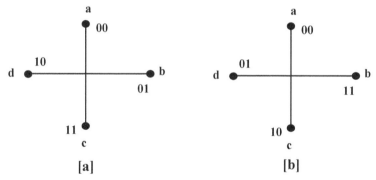

Figure 5.8: Bit Allocation to Constellation Points

An error occurs when added noise causes an incorrect constellation point to be chosen over the one actually sent. Clearly, from the constellation diagram, adjacent points are more likely to be chosen in error than are those diagonally opposite. Thus, if **a** is sent, points **b** and **d** are more likely to be chosen in error than is point **c**. In Figure 5.8.a, the two binary digits are allocated so that those on adjacent constellation points differ by only one digit. In Figure 5.8.b, they can differ by two digits. Consequently, at reasonable error rates, the former arrangement will give mainly single bit errors, and these can be corrected by the forward error correction. In the other arrangement, about half the error events produce two-bit errors and cannot be corrected.

This simple example illustrates an important concept. In keeping with the terminology "Hamming Distance" used for bit sequences, the geometrical distance between constellation points on the constellation diagrams is often called the "Euclidean Distance." The art in marrying modulation and forward error correction schemes is to associate bit sequences having the smaller Hamming Distances with the modulation points having the greater Euclidean Distances. This is not always as straightforward as it might seem. We will now show how it has been achieved for convolutional coding in the technique known as Trellis Coded Modulation. For ease of explanation, we will consider an 8-PSK system with 2/3 rate coding.

The reader may wonder why we use 8-PSK for this example, and not 8-VSB since the ATSC system does employ trellis coded modulation. It is simply that the explanation is easier to understand with 8-PSK. As the discussion proceeds, it will become clear that some of the measures used to differentiate between signal positions on constellation diagrams are more straightforward in the PSK system. The governing principles, however, are the same for all modulation schemes.

Two-thirds rate coding often adopts the principle of splitting the incoming bit stream in two, encoding one resulting stream in a rate one-half encoder and leaving the other stream unencoded. This effectively generates three output bits for each two incoming. It readily lends itself to eight-level modulation, or to be more precise, to a modulation scheme with eight positions in its constellation diagram. The arrangement, using our previously discussed encoder is illustrated in Figure 5.9.

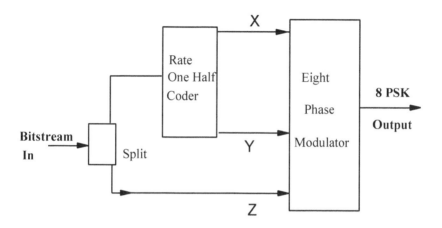

Figure 5.9: Trellis Coded Modulation

The receiver must operate on a trellis diagram similar to that previously shown in Figure 5.4. Notice that changes of state were triggered by four events, namely the arrival of any of the bit pairs 00, 01, 10, or 11. In the expanded scheme, each transmitted symbol now represents three bits, not two, and there are eight positions on the constellation diagram. Hence, we represent each of our four triggering events by a pair of constellation points. For example, one event might be described as "a change from position six to position eight" and so on. The trick is to define these four events in such a way that those that trigger similar paths through the trellis diagram have large Euclidean Distances. Figure 5.10 will be used in the explanation of how this is done.

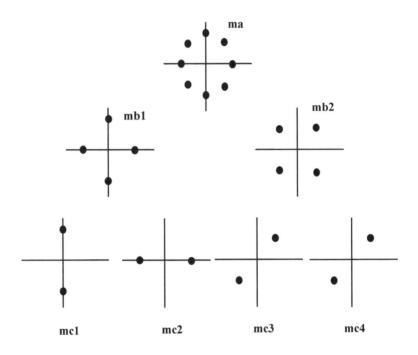

Figure 5.10: Modulation Constellation Decomposition

The eight constellation points are resolved into four pairs of two, designated **mc1, mc2, mc3,** and **mc4**—**mc** standing for "modulation condition." Notice that the points within each pair are separated by the greatest available Euclidean Distance.

For our example trellis encoder considered previously, Figure 5.8 showed some possible paths through the trellis that the receiver might choose in the event of an error. This is redrawn in Figure 5.11, for the expanded system, now with the branches relabeled with the **mc** designations. Hence, as before it is possible now to trace the possible error paths through the system, like the path **abda** for instance. We can then assign to each path the Euclidean Distance by which its two **mc** conditions differ from the correct sequence, and so calculate the probability of the erroneous path being preferred. Notice, however, that the principle protection against this type of error is the large Hamming Distance between the erroneous and correct paths, as it was for the simpler system considered previously.

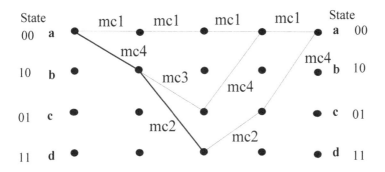

Figure 5.11: Modulation States Traversed

In addition to the receiver taking an erroneous path, there is now the additional possibility that one point in an **mc** constellation pair could be mistakenly identified as the other point in the same **mc** pair. In that case, the receiver would continue along the **aaa** path, and the Viterbi algorithm would not detect this error. However, the two constellation points in question are separated by the maximum possible Euclidean Distance.

This second type of error nearly always dominates. Notice, however, that this system, whilst using eight point modulation, is effectively transmitting only two bits per symbol due to the redundancy introduced by coding. A 4-PSK unencoded system could transmit at the same rate. However, for the principal error mechanism, the Euclidean Distance is √2 times as great as that in the 4-PSK system. Hence, the Trellis Coded system has a power advantage of 3 dB over an unencoded system. The term *coding gain* is often used to refer to this 3 dB advantage.

Applications

The coding scheme used in the ATSC standard is illustrated in Figure 5.12.

Figure 5.12: Coding Used in ATSC Standard

This uses Trellis Coded Modulation similar to that just described. Two input bits are coded into the three bits required to define one of the eight symbol amplitudes. This is done by leaving one bit unencoded and passing the other through a simple rate one-half coder. As mentioned in Chapter 8 on ATSC, there are in fact 12 of these encoders and each operates on data bits spaced 12 apart in the payload stream.

With eight-level modulation like this, as distinct from eight phase modulation, it is conceptually more difficult to discern the meaning of Euclidean Distance, since the error susceptibility of the two outmost levels differs from that of the inner levels. Hence, this system was not employed in the previous explanation of Trellis Coded Modulation. Nevertheless, the principle is the same.

The encoder applied in the European DVB standards is illustrated in Figure 5.13.

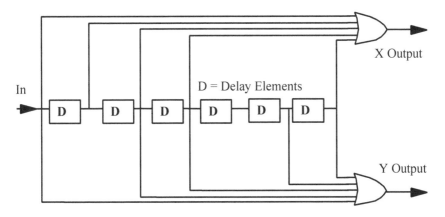

Figure 5.13: Encoder Used in DVB Standards

This encoder is capable of being adapted to various coding rates by controlling the switching between the X and Y output channel. For rate one-half the transmitted sequence is XY for each input bit as in our simple example. For rate three-quarters, the output is the four-bit sequence XYYX for every three bits input.

FEC in Hierarchical Systems

Hierarchical modulation generally refers to the technique whereby some selected bits in the transport stream are made more robust than others to transmission errors. The treatment in Chapter 2 showed how this can be done during modulation by manipulating the manner in which these bits were assigned positions on the constellation diagram. Effectively, the bits were given a power advantage over others. Now, in addition, we can show how the priority bits can also be assigned a more robust forward error correction scheme and interleaving process. A common arrangement is that shown in Figure 5.14.

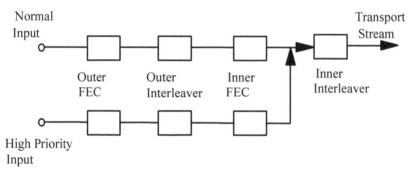

Figure 5.14: FEC in Hierarchical Transmission

With this scheme, the normal and high priority bits are given different inner and outer FEC's and different outer interleaving. Once combined into the same transport stream they share the same inner interleaver. The transport stream is then applied to a hierarchical modulator as described in Chapter 2. This is the arrangement appearing in current broadcasting standards but, of course, many variations on a theme are possible, and probably do appear in other systems.

Chapter 6

Audio Compression and Coding

The compact audio disk is often cited as the prime example of how adopting digital techniques can bring better and cheaper solutions than analog methods have ever done. Yet despite the fact that the CD almost entirely replaced vinyl records many years ago, digital techniques have been slow to penetrate into transmission systems for broadcast audio, either for radio services themselves or for the audio channels associated with video in TV. One of the reasons for this is that digital systems delivering high audio quality, comparable to the quality of FM radio, have required relatively enormous bit rates prior to the development of the compression techniques described in this chapter.

Essentially, digital audio transmission entails sampling the signal at a certain rate, approximating the sampled values with numbers, and then transmitting that series of numbers. The rate at which the samples must be taken is related to the audio bandwidth by the sampling theorem. Since each sample value must be approximated from a finite range of numbers there is a small error for most samples. The sequence of these small errors in the transmitted data stream gives rise to what is known as quantizing noise in the reproduced signal. This chapter outlines the principles underlying the sampling theorem and quantizing noise, and shows that they both contribute to the high bit rates required for good-quality audio.

In recent years, a series of experiments has shown that the human auditory system does not necessarily use all the detailed structure of the audio signal that these high bit rates are capable of conveying. In other words, much of the information is actually redundant and can be discarded with no effect on the perceived sound quality. The redundant information is removed by using mathematical models of human hearing to control the processes of sampling and quantization. These are known as psychoacoustic models. In broadcasting, the most common applications are found in MPEG and Dolby AC-3 compression systems, which are briefly described in this chapter.

Bandwidth and Sampling

In its simplest form, the sampling theorem states that for accurate reproduction a signal of bandwidth B hertz must be sampled at a minimum rate of 2B per second. To demonstrate this, let us consider the simple sampled transmission system shown in Figure 6.1.

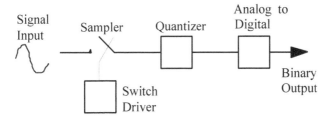

Figure 6.1: Sampled Signal Transmission System

The sampler on the left-hand side takes samples at time intervals T. The samples are quantized and sent to the receiver through some transmission system as a sequence of numbers. The numbers might be transmitted using any of the methods described in Chapter 2. At the receiver, the sequence of numbers is converted back into a stream of pulses that have the same quantified amplitudes as they had at the transmitter. These pulses are then passed through a low-pass filter of bandwidth B to approximate the original signal. This is a simple, but accurate, model of sampled digital transmission.

Now instead of thinking of the signal being sampled at the transmitter, consider instead that the transmitter generates a stream of very narrow pulses at periods of T and with amplitude unity (one), and then that this stream is multiplied by the audio signal. The resulting amplitude of each pulse is the same as it would have been under the sampling regime, but the multiplication allows us to introduce the notion of modulation, familiar to all who deal with radio or broadcasting transmitters.

The unmodulated pulse stream is periodic with period T. It follows from elementary Fourier series theory that the spectrum is comblike with frequency components at multiples of 1/T. Now, if each component is regarded as a carrier, multiplication by the audio signal constitutes a modulation of each carrier by that signal, just as it would in a radio transmitter. The effect on the spectrum is also the same as it would be in a transmitter, namely that each carrier becomes surrounded by sidebands reproducing the baseband spectral components in the audio waveform. These spectral relationships are shown in Figure 6.2.a.

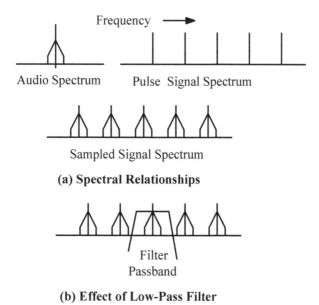

Figure 6.2: Illustration of the Sampling Theorem

It should be evident that we could reproduce the baseband signal by passing the modulated pulse stream through an ideal low-pass filter as illustrated in Figure 6.2.b. If there is to be no interference from spectral components surrounding the harmonics of the sampling frequency, then clearly the relationship $1/T > 2B$

must hold, or T < ½B, which is just the requirement expressed in the sampling theorem.

One should be careful to distinguish the spectrum of the sampled signal described from the spectrum we outlined in Chapter 2 for the digital transmission channel. Recall that the samples themselves are not transmitted. Rather the sample values are converted into binary numbers and the numbers are transmitted as a sequence of bytes, which is totally different from transmitting the samples themselves as voltage waveforms.

Now turning to quantizing noise, suppose each sample is to be transmitted as a number represented by four bits, which means that it would be quantified into 15 levels, say the voltages +/-7, +/-6, . ., +/-1, 0, to keep things simple. Thus if the actual signal level were 4.8 V, this would be transmitted as 5 V, giving an error of 0.2 V. If a noise voltage of 0.2 V had been added to the original signal before quantization, it would have had the same effect as this error, and hence the concept of quantizing noise arises. Naturally, real systems quantize the samples to far more than 15 discrete levels and use far more than four bits per sample.

The quantizing noise level can be calculated by statistical techniques, but in practice this is fairly difficult. Uniform quantizing levels as described are rarely used. Fine steps are used at low signal levels and larger steps at higher levels in a process known as *companding*, and reflecting the probability that the signal spends more time at the lower levels. Thus the more probable errors are kept small, and the errors likely to occur less often are allowed to be larger. These characteristics are embodied in companding laws that have been standardized in many applications. Thus one will see references to A-law or C-law companding.

We can now look briefly at how these factors have controlled the bit rate necessary for high-quality audio transmission. It follows from the sampling theorem that to faithfully reproduce a signal with an audio bandwidth of 20 kHz, one must have a sampling frequency somewhat above 40 kHz, and 44.1 and 48 kHz are commonly found. For quantizing noise to be inaudible, it is generally held that the audio samples need to be quantified over greater than 50,000 levels requiring 16 bits per sample. Thus, most high-quality audio reproduction generally requires 2%48%16Kbits/s or 1.5 Mbits/s for a stereo channel. This is typical of the bit rates found in compact audio disk applications.

If we wished to broadcast this signal using one of the more robust modulation techniques described in Chapter 2, say quadrature phase shift keying (QPSK) at 2 bits per symbol, the bandwidth required would be about 400 kHz, which is much greater than the channel bandwidth required for an analog FM service and a substantial part of a TV channel. Hence, the need for bit rate compression is evident.

Psychoacoustic Masking

As mentioned, studies of human hearing indicate that not all the information in an audio signal is actually required for it to be perceived as high quality. Evidently some of the information is redundant and could be discarded before transmission. Just what can be discarded is determined by psychoacoustic models that mathematically model the processes of hearing and perception.

The studies used to develop the psychoacoustic models have revealed that many aspects of hearing can be explained by supposing that the ear decomposes the sound signals as if they were passed through a bank of band pass filters covering the audio spectrum. The pass bands of these hypothetical filters are known as the critical bands. They tend to be broader in bandwidth as their center frequencies increase. It is a characteristic of the critical bands that the perceived character of a stimulus remains relatively stable if its frequency components are confined to one critical band, but the character changes abruptly if it shifts into another critical band or if the spectrum spreads into adjacent bands. Thus the critical bands are relatively independent channels on which the psychoacoustic models may operate.

The redundancy in hearing is manifested in the way that under certain conditions the signals in some critical bands may not be audible and therefore may not need to be reproduced. In particular, a high-level signal in one critical band may mask lower-level signals in adjacent bands if they fall below some threshold determined by the level of the dominant signal. This masking occurs while the louder signal is present, but also persists for brief intervals before and after the louder sound occurs. In addition, a dominant tone can mask noise in its own critical band.

Broadly speaking, this masking effect is exploited in audio compression by sampling the signal level in each of the critical bands and dynamically assigning a bit rate to each according to the different resolution required by each over short periods. If a critical band contains a high-level tone, for example, it can mask quantizing noise in its own and adjacent critical channels. Hence, the quantizing noise in these bands can be relatively high. Additionally, if the adjacent bands contain signals low enough to be masked themselves, it may not be necessary to send data for them at all. The psychoacoustic model embedded in logic circuitry in the coder determines what level of quantizing noise is acceptable in each band. It adjusts the step size by which the sample is approximated by applying a scale factor to each band and then allocates a proportionate share of the total bit rate to the band. Figure 6.3 illustrates the concept of critical bands and masking. In this particular case, the figure shows 32 sub-bands, but as we shall see that number varies from implementation to implementation.

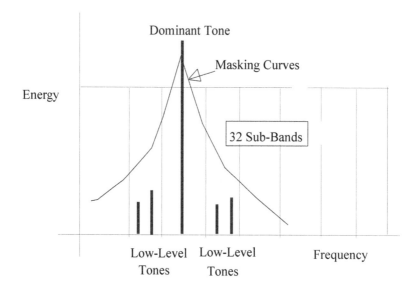

Figure 6.3: Principle of Psychoacoustic Masking

The filter bank defining the critical bands does not normally exist in the coder. Rather, the coder operates mathematically on the spectrum found by taking a digital Fourier transform (DFT) of the incoming signal over a short period of time. The data applying to the signal over each short period are assembled into a package of data, known as a frame. In addition to the quantized sample values, each frame carries a digital code conveying to the receiver the bit allocation and the scale factor applying to each critical band.

The frame normally carries data for more than one channel, for example a stereo pair, but frequently there are up to six channels comprising surround sound channels or even separate language soundtracks. A typical frame might be made up as follows:

- The header, indicating the source.
- A burst of bits comprising the cyclic redundancy code for forward error correction.
- The bit allocation table, showing how many bits are allocated to the output of each filter in the filter bank.
- The scale factor table, showing what scale factor (multiplier) is applied to the output of each filter.
- The sub-band samples—the digitized level in each sub band.

The sequence of frames, transmitted in time order, comprises what is called the packetized elementary stream for the audio service. The concept of packetized elementary streams and the way they are assembled together to form the transport stream for a complete service, are described in Chapter 4.

Practical Implementations

The most common implementations of this type of compression in broadcasting are Dolby AC-3 and the audio components of the MPEG Standards. Dolby AC-3 has been adopted for the ATSC Standard in the United States. MPEG-1 is an early implementation for coding mono or two channel stereo services, while MPEG-2 has a multichannel capability. The MPEG-2 specifications encompass three different *Layers*. Layer one implements the simplest psychoacoustic model and provides a modest degree of compression, while Layer three is much more complex and intended for applications where low bit rates are critical. Layer two is intermediate between the two. In broadcasting, Layer two is normally applied to program transmission, but Layer three may be found in critical program distribution, such as in the use of ISDN contribution circuits.

Dolby AC-3 analyzes each channel effectively over 256 audio samples, giving a time resolution of 5.3 mS (at 48 kHz sampling rate) over which the signal statistics are assumed constant. However, this resolution can be halved dynamically if transients are sensed by the coder circuitry. The filter bank has a uniform resolution of 93 Hz, from which the psychoacoustic model determines the critical band occupancy. On initialization the coder and decoder assume a set of core values for the model, and these are changed as necessary by the transmission of side information in the frame.

MPEG Layer 2 operates over a block of 1152 samples giving a time resolution of 24 mS (at 48 kHz sampling rate). The processing emulates a set of 32 filters in the filter bank. Information on the scaling factors and bit allocations in each band is transmitted to the decoder in every frame.

Both standard systems are capable of multichannel operation on a variety of sampling rates. Both are capable of high degrees of compression. It is said that most people find the quality of a stereo service indistinguishable from the original at a bit rate of about 128 Kbits/s. This is a reduction by more than a factor of 10 over the uncompressed bit rate.

The latest development from MPEG is MPEG-2 AAC, where AAC stands for *advanced audio coding*. This exhibits a much greater degree of compression than Layer two and is finding application in the emerging systems for IBOC digital radio standards. AAC employs a much more complex filter bank than does Level two and adds a number of features called *tools* in the MPEG terminology. Two of these, in particular, are the use of prediction methods and Huffman coding within the output of each filter in the filter bank. These techniques are essentially the same as those employed for video compression, and the reader will find a description of how they work in Chapter 7 on that subject. The standard describes three levels of complexity, which may be selected according to the bandwidth or bitrate available and, of course, to the affordability of receivers. The latter is important in international HF broadcasting. For high-quality applications, the necessary bitrate for a given audio quality is said to be half that required of Layer two.

Chapter 7

Video Compression: The Techniques Behind MPEG

As explained earlier, the acronym MPEG stands for Moving Pictures Expert Group, which is formally a part of a joint technical committee in the International Standards Organization (ISO) and the International Electrotechnical Committee (IEC). MPEG has defined techniques for compressing the data rate required to transmit video services in digital systems. In broadcasting, these standards are used in the ATSC system in the United States and in the European DVB systems.

The initial specifications commonly referred to as MPEG-1, had reasonably modest objectives, being intended for low-resolution pictures such as computer generated images transmitted or stored at data rates up to 1.5 Mbits/s. MPEG-2, which essentially was completed only in 1994, extends the capacity to full broadcast standard video. MPEG-1 and MPEG-2 share many techniques in common to exploit the redundancies found in typical video sequences. Adjacent parts of a scene are very similar to each other, as are successive frames in the sequence of pictures that comprises the motion picture format. Where data to be transmitted change slowly, as with these characteristics, it is intuitively obvious that transmitting the changes is much more economic than transmitting everything and thereby repeating the static elements over and over. This is the common principle behind differential pulse code modulation (DPCM) systems. We shall show how these systems have been cleverly adapted by MPEG. Additionally, MPEG systems do not work on the picture elements (the pixels) themselves, but mathematically transform the pixels values using the discrete cosine transform or DCT. This transformation gives a stronger correlation between adjacent samples and many of the values generated can be discarded because their absence tends not to be noticed by the human visual system in the reconstructed images. The first property of the DCT makes the DPCM process much more effective, and the second substantially reduces the number of values requiring transmission.

Having defined picture encoding using these techniques, the MPEG standards next address the coding of the data for transmission. They provide for two common methods of data compression on the transmission link. These are run length coding and Huffman coding, the latter sometimes being described as entropy encoding.

All these processes comprise what is commonly called the "MPEG toolbox," a fitting name in that the standards allow a system designer to choose some or all of them and the extent to which they are applied in any application. The system is very flexible and effective. A few years ago, the standard transmission rate for broadcast quality video distribution was regarded as 144 Mbits/s. Some users came to accept 34 Mbits/s with the development of early compression systems, and now 8, 6, or even 5 Mbits/s is generally considered adequate with MPEG-2. Let us now look at some of the elements of the standards.

Exploiting Differential Coding

Although MPEG does not use DPCM as such, the philosophy behind DPCM is very useful to develop an understanding of the MPEG techniques. Consider the simple DPCM system shown in Figure 7.1 and suppose it is transmitting some signal where the value of each successive symbol depends strongly on that preceding it, as happens for example with the pixels in a video signal. This system, as drawn, transmits only the changes between symbols, and the actual symbol values are recovered by the circuitry around the delay unit in the receiver, whereby the previously derived value is systematically added to the change value as it is received. The reader is invited to plug in a simple sequence of numbers to see that this rudimentary arrangement really does work.

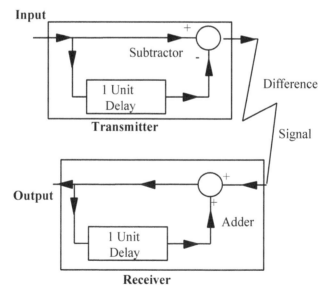

Figure 7.1: Rudimentary Differential Transmission System

In Figure 7.1, the function of the delay line in the transmitter is essentially to "predict" the value of the next symbol to be transmitted. If this prediction is reasonably accurate, then the DPCM system needs to transmit only the errors and the numerical range of the errors should be much smaller than that of the original symbols, leading to a considerable reduction in the data rate. The system as drawn uses only the value of the previous symbol as the prediction, but obviously if we could exploit some stronger property of the data stream to form a better estimate, the range of the errors and the data rate could be reduced even further. This is the essential principle of MPEG. On this principle, we can replace the simple delay units in Figure 7.1 with "predictors" and arrive at the system shown in Figure 7.2.

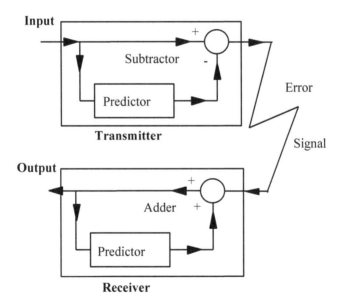

Figure 7.2: Elementary System with Prediction

Ideally, this system can work, but it suffers from the very serious deficiency that the predictors in the transmitter and receiver operate on essentially different signals. This makes designing the two predictors to track each other accurately very difficult, particularly when some processing of the transmitted signal, like quantizing, is introduced. Fortunately, there is a way around this by embedding the predictor in a feedback loop at the transmitter, as shown in Figure 7.3. This figure shows a quantizer in the transmission path to more accurately represent a real system and to illustrate the point that the two predictors here operate on the same signal.

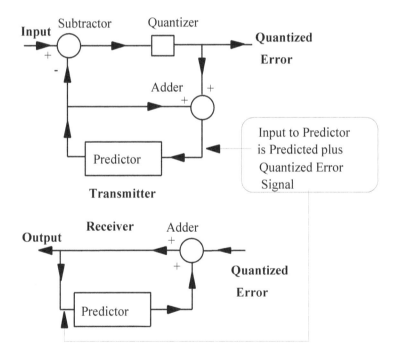

Figure 7.3: DPCM Transmission System

Figure 7.3 is the essence of the MPEG picture encoding system. The better we can make the predictors, the lower the data rate we have to transmit. The very clever part of the MPEG standards is in the prediction algorithms, but before addressing that, let us revert to the two coding methods mentioned, run length and Huffman coding, as they have a bearing on how the prediction algorithms have been chosen.

Run Length Coding

In some classes of data, particularly those arising from video pictures or facsimile scans, one often finds long strings of numbers with a constant value, especially zero. It is more efficient to transmit this as only two numbers, the constant value and the number of symbols in the run, rather than as the full sequence of the same number. This is known as *run length coding* or more often *variable run length coding* (VRLC). If one can transform the raw data in some fashion to make long strings occur often, then a reduction in the required transmission rate may well result. We shall see that MPEG achieves this by way of the DCT mentioned.

Huffman Coding

Huffman Coding, named after its inventor, is an example of entropy coding, and is simple in principle. Each symbol (or value) to be transmitted is assigned a binary code word of some number of bits. Words with a small number of bits are assigned to those symbols that occur more frequently, and longer words are assigned to symbols that occur only rarely. Thus, for example, fewer bits need be transmitted than in a system where all symbols are assigned words of equal length. A well-known example of the principle occurs in the Morse code. Huffman invented a systematic way of assigning words to values that has been adopted in the MPEG standards.

Huffman and run length coding are quite standard techniques in all classes of data transmission, and they are "lossless" in that, barring transmission errors, the original data can be regenerated exactly. This is not the case with the other techniques used in MPEG, in which the data (for example, pixel brightness) may be reproduced only approximately by the predictors in the transmitter and receiver. The art of MPEG has been not only to make this "lossy" data reconstruction subjectively acceptable, but also to format the transmission parameters to make the lossless techniques particularly effective. The DCT, which we now address, is an important element in this process.

The Discrete Cosine Transform

The discrete cosine transform is very closely associated with the Fourier transform, and like the latter, may be considered to extract the "frequency components" in a sequence of samples. In MPEG, the DCT operates over blocks of eight rows by eight columns of pixels in the video picture, giving 64 numbers to be transformed in each block. The rows are runs of eight pixels taken from each of eight successive scanning lines and on the picture screen each block comprises a small square area of adjacent pixels. The process is illustrated in Figure 7.4, which purports to show a block of pixels transformed into a similar block of DCT coefficients. For simplicity, Figure 7.4 shows only 4 × 4 blocks. In a sense, the DCT coefficients across the block are a measure of the frequency components in a horizontal scan of the picture, and those down the block a measure of frequency components in a vertical scan. Higher-frequency components occur toward the right and bottom of the block and have large values only in busy scenes with a lot of detail. These values approach zero in scenes with little changing detail, such as pictures of the sky, for example. The overall average brightness of the pixel block is contained in the upper left-hand corner.

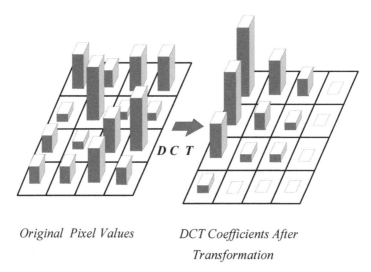

Original Pixel Values   DCT Coefficients After
Transformation

Figure 7.4: Discrete Cosine Transform

Figure 7.4 attempts to show two common properties of the DCT that occur in most video applications. Normally the higher-frequency components diminish rapidly in value away from the top left corner, and in addition the correlation between values in adjacent cells is more marked than with the original pixel block. In viewing a reconstructed picture, the eye is increasingly less sensitive to the accuracy of the DCT coefficients toward the right and bottom of the DCT block, and these can be coarsely quantized or set to zero, according to the quality of reproduction desired. On the other hand, the coefficients toward the upper left corner govern the overall brightness of each reconstituted block of pixels and must be quantized finely and transmitted accurately. Otherwise, the brightness in the reconstructed scene tends to change abruptly from block to block leading to a checkerboard effect commonly called "blockiness."

To exploit these properties further, MPEG uses a scanning sequence in each block like that illustrated in Figure 7.5.b, rather than what might be considered the more natural sequence in Figure 7.5.a. Notice that this gives a smooth transition through adjacent values, avoiding large jumps in magnitude, and secondly, it can generate long runs of zeros or constant values where the higher-frequency DCT coefficients are coarsely quantized or set to zero. The first property favors efficiency in the DPCM process, as explained, and the second neatly fits in with the variable run length coding used on the transmission link, as mentioned.

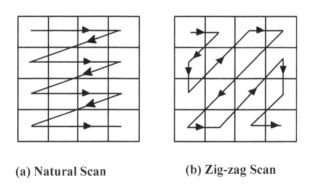

(a) Natural Scan   (b) Zig-zag Scan

Figure 7.5: The Zig-zag Scanning Sequence

The relationship between picture quality and the accuracy or transmission of the higher-order DCT coefficients adds another degree of flexibility in MPEG. It should be evident that in the inherent DPCM process, the prediction error will be greater in rapidly changing scenes and would ideally be transmitted at a higher data rate than otherwise. However, this is generally difficult in a world where most digital links have a fixed rate. With MPEG, the outgoing data are stored in a buffer that is emptied at the fixed rate but can be filled at a varying rate. If the buffer tends to overflow, a feedback loop into the picture coder causes the quantizing of the higher-order DCT coefficients to be made coarser, decreasing the volume of data and increasing the prevalence of long run lengths of fixed values and favoring the variable run length coding. Thus, normal picture sequences are encoded to the defined quality, but this quality declines for "busy" sequences or those with rapid motion. It is in just these circumstances, however, that the eye is least sensitive to picture quality. In a sense, MPEG achieves the desirable aim of matching its coding to its channel, of which the human visual system is part.

Motion Prediction

We have seen that the major essence of MPEG systems, arising from the principles of DPCM, is that they should transmit only the differences between a current block of data and a predicted block of data derived from what has gone before. The smaller this difference, the smaller the required data rate. In other than still pictures, the blocks of data will change position from frame to frame, and in predicting the values of a particular block, the MPEG predictor could very usefully use information on where the block came from in previous frames. Effectively, the predictor should keep track of moving blocks, as they change relatively slowly within themselves, compared to how quickly a block in a fixed position on the screen would change when motion is present. Think, for example, of the situation of a camera scanning a crowd scene. If we could track an individual face through the sequence of picture frames, the pixel blocks within that face might stay essentially unchanged in their values.

To give effect to this technique, MPEG systems transmit with each DCT block a motion vector indicating the velocity of the block across the screen in the horizontal and vertical directions. This tells the predictor in the receiver where the block will be next, or perhaps we could better describe this as the receiver knowing where the block came from to determine what its values were last time. There are various ways of estimating the motion vector, but most involve the transmitters moving the 8×8 block around a 16×16 "macroblock" in adjacent picture frames to find the position of best fit. This process, incidentally, is not defined by MPEG, which confines itself to requirements at the receiving end. Hence, the standards leave room for innovative solutions.

The MPEG Frame Structure

As described, the DPCM process embedded in MPEG has two problems. Essentially, it has to start somewhere—we cannot have a cascade of picture frames all derived from those that came before and not have a beginning. Second, transmission errors will propagate and affect all frames that come after the one in which the error occurs. In addition, in video, there must be frames capable of standing alone to permit editing, cutting, and splicing of program segments.

To satisfy these considerations, MPEG defines three types of frames to be transmitted. The first, called intra-frames or I frames, are stand-alone and independent of any previously transmitted frames. They are encoded entirely from within themselves and achieve only a moderate degree of compression. The second type comprises the predictive frames or P frames. These are coded using motion compensated prediction from previous I and P frames. They achieve considerably more compression than the I frames. Finally, there are the bidirectionally predicted or B frames. These are derived from past and future I and P frames and are not used for further predictions. They achieve the highest degree of compression.

The diagram in Figure 7.6 shows a typical sequence of frames and the interdependencies between them. Note that the frames are not transmitted in this order, as it considerably simplifies the receiver design if the P and B frames follow those on which they depend. The boxed material in Figure 7.6 shows the sequence in which the frames would actually be transmitted. (Note that this is a little simplified for clarity. In practice there would be B frames between P7 and I8 backwardly dependent on I8.) The out-of-sequence transmission does require frame storage in the receiver, and the standards do define a "simple profile" which omits the B frames.

Although it does not concern us so much in transmission, the out of sequence nature of the MPEG frames does introduce significant complications in video editing. Equipment capable of cutting and inserting pictures in compressed format has become available only comparatively recently. Prior to that it was necessary to convert the video back to its native PAL or NTSC format for editing.

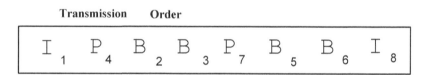

Figure 7.6: Frame Dependencies and Transmission Order

The MPEG standards define "profiles" and "levels." The profiles essentially describe the features of the toolbox that are included in a particular system, including some features not yet finalized such as scaling the fineness of quantizing according to the signal-to-noise ratio. The level largely defines the input source parameters, such as the horizontal and vertical resolution of the picture and the maximum sampling rate. The following tables give an abbreviated indication of the parameters encompassed within the most common profiles and levels.

109

Profiles	
Simple	I and P Pictures 4:2:0 Sampling
Main	I, P, and B Pictures 4:2:0 Sampling
High	I, P, and B Pictures 4:2:2 Sampling

Levels	
Low	325 × 288 Pixels 30 Frames/sec 4 Mbits/s
Main	720 × 576 Pixels 15 Mbits/s
High	1440 × 1152 Pixels 60 Frames/sec 80 Mbits/s

The bit rates quoted are the upper bounds, and lower values may be found in particular applications. At present, readers are most likely to encounter the main profile, main level system suited to standard definition NTSC or PAL pictures.

Video Sampling

For the benefit of readers who may not be familiar with the nomenclature of video sampling, this note may be helpful and give a little interesting background.

In Chapter 6 on audio coding and compression, we showed that any signal having a bandwidth B must be sampled at a rate somewhat greater than 2B if it is to be reproduced faithfully. This applies to video as much as to audio, and in particular to the luminance and the two-color difference signals of component video. In practically all the world standards for analog broadcast video, the luminance channel of standard definition video is assigned a bandwidth of around five megahertz. Thus, one would expect the sampling frequency of the luminance channel to be somewhat greater than 10 MHz. The color difference channels are usually assigned smaller bandwidths, but not in every application, and hence the sampling rate should generally, but not always, be lower.

In the early days of setting standards for digital video, these considerations indicated that a convenient rate for sampling the luminance could be four times the color sub-carrier frequency for the system in use, i.e., at about 14.4 MHz for most NTSC and at about 17.3 MHz for most PAL systems. For internal studio use, the color difference signals were to be sampled at the same rate, but half that rate was proposed for contribution circuits and other transmissions outside the studio. Because of the relationship between the sampling frequency and the color sub-carrier frequency, these two schemes became known as 4:4:4 and 4:2:2 sampling respectively. A third scheme was also proposed for situations where lower quality could be tolerated and the available bandwidth might be limited. This involved sampling the color difference signal on only every second scan line, and on reproduction, filling in the missing color information by extrapolation between lines. Not surprisingly, this scheme became known as 4:2:0 sampling.

It soon came to be realized that having different bit rates for different video systems would be a great nuisance for transmission over national and international communication circuits carrying telephony and much else as well. Chapter 4 in this book describes how digital traffic must be assembled into transport streams. Having to deal with a variety of bit rates greatly complicates this process. Consequently, the international standards for digital video now specify 13.5 MHz for the luminance component sampling in all systems, with the color difference sampling rates adjusted accordingly. Nevertheless the 4:x:x convention still exists. In it the "4" indicates 13.5 MHz sampling for standard definition video, but may be much higher in high definition applications.

The standard referred to here was originally promulgated by the ITU as CCIR Recommendation 601, so that one still sees video material quite often referred to as "Rec. 601" quality. A little mental arithmetic shows that a 4:2:2 picture at 8 bits per sample requires a raw transmission rate of around 250 Mbits/s, so the need for compression is quite obvious.

Chapter 8

ATSC 8-VSB System

As mentioned in Chapter 2, the ATSC 8-VSB modulation system adopted in the United States for advanced digital TV is derived directly from linear modulation. Consider a baseband stream of digital symbols that has the form and spectral characteristics described in Chapter 2 as being necessary for band-limited digital transmission. If we were to apply direct linear double sideband modulation to this stream, the result would be a robust digital communication system. However, as in analog NTSC or PAL transmission, the bandwidth that this signal would occupy is twice that strictly necessary. As in the analog case, this arises because both the upper and lower sidebands carry the same information. Hence, one set of sidebands can be discarded without losing any information. In analog TV, the use of vestigial sideband (VSB) modulation has been the standard method for many years. The ATSC standard now extends this to the digital domain.

In this chapter, we look at some of the features of the ATSC system and show how they arise from the principles of digital transmission outlined in earlier chapters. The ATSC has adopted eight symbol levels for terrestrial transmission and 16 levels for cable systems, where the environment is more benign in terms of interference and impulse noise. The discussion focuses mostly on the eight-level system as the eight-level and 16-level systems are very similar in principle.

Symbol Characteristics and Spectrum Shaping

It is instructive to consider the characteristics of an eight-level baseband system converted first to a double sideband (DSB) system by direct linear modulation, and then to consider what is required to convert this into a VSB system. Ultimately, the system must occupy a standard 6 MHz TV channel, implying that the DSB version can have a spectrum width of approximately 12 MHz. The ideal rectangular spectrum would be like that shown in Figure 8.1.a, but remember from the discussion in Chapter 2 that we must have a symmetrical rolloff at the channel edges according to the Nyquist criterion introduced in that chapter. The Nyquist rolloffs are shown superimposed on the ideal in that figure, and we see that the ideal rectangular spectrum must occupy something less than 12 MHz, or equivalently, the ideal baseband channel filter must encompass a little less than 6 MHz. In fact, the ATSC has chosen a Nyquist rolloff of 0.31 MHz and an ideal filter bandwidth of 5.38 MHz. From Chapter 2, we see that this gives a symbol rate of 10.76 megasymbols per second. There is a compromise here, as a wider Nyquist rolloff would have given a sharper symbol less extended over time with less potential intersymbol interference, but it would also have lowered the achievable symbol rate. This choice keeps the symbol rate close to the maximum achievable.

(a) Ideal Rectangular Spectrum

(b) Spectrum with Nyquist Rolloffs

Figure 8.1: Spectral Relationships in VSB

In converting this system to VSB, any vestigial slope characteristic that is symmetric about the carrier frequency would serve, even that shown in Figure 8.1.b, which one often sees in discussions of analog VSB. However, it makes sense to keep the filter characteristic symmetrical about the center of the transmission channel as practical filters that may be used to realize this characteristic in intermediate frequency stages all have that property. Accordingly, the VSB region has been defined as having the same form as the Nyquist rolloff at the top end of the channel. One sees that this puts the very bottom of the spectrum at 0.31 MHz below the carrier frequency, or in other words the carrier must be 0.31 MHz above the bottom edge of the channel.

115

Taking all this into account, we see that the 8-VSB channel characteristics are defined as in Figure 8.2. Actually, this figure shows the overall channel characteristic defined by the combination of filtering at the transmitter and at the receiver. We have shown that for an optimum noise immunity the filtering should be shared equally between the transmitter and receiver. Thus, for a transmitter or receiver alone, the filter characteristic will have values that are actually the square root of those shown in the figure.

Figure 8.2: Practical VSB Channel Characteristic

Figure 8.2 shows the pilot signal at the carrier frequency, as required by the ATSC standards. The pilot is required as a frequency and phase reference for coherent demodulation at the receiver. Unlike their analog TV predecessors, digital VSB systems cannot be demodulated with an envelope detector and coherent demodulation is always necessary. The reason lies in the baseband symbol stream in which the symbol amplitudes can take on the normalized values +/-7, +/-5, +/-3, and +/-1 V. Note that these include negative values. By design, modulated analog TV signals are never allowed to go negative. A negative value implies a phase shift of 180 degrees in the carrier, and this cannot be detected by an envelope demodulator. Hence, coherent demodulation is required because of the baseband characteristics of the digital symbol stream.

It should now be apparent that the parameters defined for the VSB channel in the ATSC standards have their genesis in the simple principles of basic baseband digital transmission systems.

ATSC Framing Structure

The framing structure of the transmitted signal is an important aspect of the ATSC standard. As we shall see, the framing structure not only accommodates the transport stream requirements outlined in Chapter 4, but also plays an important role in mitigating RF propagation effects like multipath and impulse noise.

We mentioned in Chapter 4 that the basic transport stream packet for ATSC consisted of 188 bytes, including a synch byte. At the transmitter, this is altered in two ways. First, the synch byte is stripped off, leaving 187 bytes to be transmitted. Then 20 bytes are added to this for the Reed-Solomon error correction, giving 207 bytes to be transmitted in each packet, which at eight bits per byte amounts to 1656 bits. The trellis coding at rate two-thirds increases this to 2484 bits, or 828 symbols, since eight-level coding gives three bits per symbol. A special waveform, known as the data segment synch, is added to the head of this packet and occupies four normal symbol periods. Thus, in total, the modified transmission stream packet now occupies 832 symbol periods, or a total time of 77.3 usec at the symbol rate of 10.76 megasymbols per second. Each packet is now called a *data segment*.

The form of the data segments is shown in Figure 8.3, which also shows the shape of the data segment synch waveform.

Figure 8.3: Form of Data Segments

Periodically, at intervals of 313 packets or 24.2 msec, a special data segment known as a field synch is inserted. The field synch carries specific data used by the adaptive equalizer in the receiver to estimate what echoes may be present because of multipath. The form of this data signal is described later in this chapter when we consider the operation of the adaptive equalizer.

Figure 8.4 shows the overall framing structure. This diagram is formed by lining up the starting time of sequential data segments at the left-hand edge, starting at the beginning of a frame synch and continuing downwards over two complete frame periods. Thus, for example, the narrow rectangle shown on the left represents the sequential data synch segments lined up one under the other. The data segments themselves can be imagined as filling in the larger blank rectangles, again lined up one under the other. Diagrams like this are a quite common way of illustrating the framing structure of some digital transport streams.

Figure 8.4: Overall Framing Structure

Securing Noise Immunity

The use of the data segment synch waveform adds greatly to the robustness of the system against impulse noise in particular. Recall that one of the worst aspects of impulse noise in digital transmission over RF channels is that it may cause the system to lose synchronization and significant time might elapse before the system regains that synchronization. The data segment synch waveforms are easily recognizable. They have only two widely spaced levels and are unlikely to be mistaken for data even during a high-noise event. The receiver locks its master clock to them using phase-locked loop techniques similar to those described in Chapter 4 for synchronizing transport stream packages.

The net effect of all this is that although the receiver may suffer data errors during a noise event, it is much less likely to lose synch. Scattered data errors are normally corrected by the forward error correction provisions and might even have no effect on the perceived quality of the service, provided that the receiver retains synchronization.

Forward Error Correction

In addition to the immunity provided by the use of the data segment synchs, the ATSC system employs two levels of forward error correction. The first is a Reed-Solomon (RS) cyclic redundancy code, which, as mentioned, operates over the 187 data bytes and adds 20 bytes to each data segment. The second is trellis encoding at rate two-thirds. The RS code can correct up to 10 byte errors per data segment. To ensure that a single burst on noise is unlikely to cause more than that number of errors in any segment, the bytes in a sequence of segments are spread out in time by being assigned more or less randomly across the segments in the sequence. At the receiver the bytes are reassigned to their correct segments, but it is unlikely that this reassignment will put together a run of incorrect bytes because they essentially come from scattered instances in time. This is the process of data shuffling mentioned in Chapter 3.

For each two bits on the input data stream, the trellis coders generate three output bits, which conveniently constitute one symbol for the eight-level baseband coding. The two to three bit mapping is chosen to constitute trellis coded modulation as described in Chapter 5 on error correction. The system actually uses 12 coders so that the symbols produced by each are spaced 12 symbols apart in the symbol stream. Again, this adds to the immunity from impulse noise as it makes it less likely that a string of consecutive symbols presented to the decoder will contain a run of errors.

Adaptive Equalization

In Chapter 3 on the effects of RF propagation, it was asserted that no digital broadcasting system could work satisfactorily if it did not have some built-in means of countering multipath propagation. The ATSC system applies a particularly robust method employing adaptive equalizers.

The general form of an adaptive equalizer is shown in Figure 8.5.

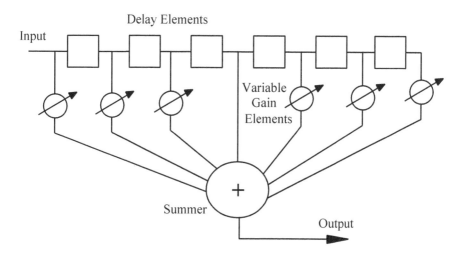

Figure 8.5: General Form of Adaptive Equalizer

It consists of a series of storage elements that hold the values taken at the receiver sampling instances over a long sequence of symbols from the stream being received. The samples move forward by one storage element as each new sample is taken

121

at the input. The main output signal is usually taken from somewhere within the chain of storage elements. Samples from the storage elements before and after the main output element can be added to the output signals at levels that may be varied electronically. Varying the levels is referred to as changing the equalizer tap values, probably because some value is "tapped off" each contributing storage element.

The net effect of the equalizer is to add leading and trailing echoes to the main signal. Hence, if these echoes can be made the negative or inverse of the echoes arising from multipath, the detrimental effects of the multipath can effectively be canceled. To do this however the equalizer must be able to determine just what multipath echoes are present. It employs a *training sequence* contained within the frame synch segment that occurs after every 312 data segments or every 24.2 msec as described.

The training sequence is a burst of pseudo-random data occupying 511 symbol periods. The term *pseudo-random* indicates that it is affected by system defects just as truly random data would be, because it contains all possible level transitions, for example. In fact, however, its sequence of values is known precisely and stored in the receiver. Thus, every 24.2 msec the receiver can subtract this stored sequence from the values currently held in the storage element chain in the equalizer. The differences reveal the magnitudes of the echoes due to multipath. From this, the necessary tap values in the equalizer can be determined to negate these echoes, and of course these settings are recalculated every 24.2 msec.

The length of the training sequence is approximately 48 usec, and in principle echoes with this delay can be compensated. This corresponds to a path length difference of nearly 15 km or 10 miles. In practice, echoes with magnitudes of up to 80% of the main signal amplitude have been successfully cancelled. That is quite an impressive performance.

Chapter 9

COFDM: The New Modulation

One of the techniques that best illustrates the advantages of digital over analog transmission is Coded Orthogonal Frequency Division Multiplexing (COFDM). The new method promises great improvements in the applications of radio propagation by doing away with multipath effects, allowing on-channel repeaters, and permitting a new service area configuration of transmitters known as the single frequency network. To illustrate its advantages consider some of the real life problems that arise in typical broadcasting situations.

For example, consider setting up links for a TV outside broadcast coverage of the running of a marathon. One could try hiring space for microwave repeaters on city buildings or using helicopters. Alternatively, COFDM would permit using reflecting signals from those buildings to create a transmission path back to the studio.

Or consider covering a large city for TV or radio broadcasting. The present practice is to seek out a very high site and install a tall tower and powerful transmitter. Alternatively, with COFDM it is possible to scatter a few small transmitters across the city, all operating on the same frequency and carrying the same program, and giving almost perfect coverage.

Or in delivering services from a satellite, the signals are often blocked into some streets by high buildings. A solution might be to put satellite dishes on some roofs, and distribute the received signals by cable. If COFDM were used, however, a small repeater, retransmitting the satellite signal on its own frequency into the shadowed areas could serve the blocked sites.

COFDM is used for terrestrial radio and TV broadcasting in Europe and some other countries, and may soon be adopted in the United States for IBOC (In Band on Channel) audio applications. In the next few years, we can expect to see it used extensively in outside broadcasting applications.

Principle of COFDM

COFDM is essentially a new method of transmitting high data rate digital signals over radio channels, or for that matter over wire and cable links. To make the numbers specific, let us look at how COFDM might be used to transmit a digital stream at a data rate of say 1 Mbits/s.

The principle of COFDM is to break this single bit-stream into a number of parallel bit-streams and to transmit all of these lower bit-rate streams on a very large number of carriers side by side within the one channel. For the 1 Mbits/s stream in this example, we may choose to break it up into, say, 500 parallel streams each carrying a bit rate of 2 kilobits per second (2 kbits/s). Each of those 2 kbits/s second streams is then modulated on to one of 500 carriers spaced 2 kHz apart. With the 500 carriers spaced at

2 kHz, the total channel occupancy in this example would be 1 MHz.

The numbers applying to the carrier spacing and the bit rate on each are chosen according to the principle of *orthogonality*. Orthogonality is a mathematical concept, but it simply means that the arrangement of the carriers is such that each can be demodulated without interference from any of the others. To provide for this, the digital symbols modulating each of the carriers are shaped so that any carrier has a spectrum of sidebands around it very similar to the form shown in Figure 9.1.a. The spectrum shows a central peak at the carrier frequency and points of zero magnitude at sideband frequencies corresponding to multiples of the symbol rate. Now if we space a number of these carriers apart so that the peak of any one falls on the spectrum zeros of all the others the overall frequency spectrum appears as shown in Figure 9.1.b.

It may be apparent that any carrier in the ensemble of Figure 9.1.b can be demodulated without interference from the others. Since the peak of any carrier coincides with the spectral zeros of all the others, the peak value is unaffected by what amplitude the others may have from moment to moment. This is the principle of orthogonality. The reader will no doubt observe that it is very similar to the method described in Chapter 2 for avoiding intersymbol interference in baseband systems. Here, we are working in the frequency domain, whereas previously the techniques applied in the time domain. This is a particular instance in which the symmetry of the Fourier transform in switching one's thinking between the time and frequency domains becomes manifest.

(a) Spectrum Around One Carrier

(b) Spectrum Around Several Carriers

Figure 9.1: Orthogonal Carrier Spectra

Most of the modulation schemes found in single-carrier systems are used without changing the essential principle. Suppose that, in our example, we had chosen to use 16 QAM on each carrier. With 16 states available, the data capacity is 4 bits per symbol. For our 1 Mbits/s example, therefore, we need to transmit 250,000 symbols per second. If we choose to use, say 1000 carriers, the modulation rate on each must be 250 Hz (strictly, 250 symbols per second), and to preserve orthogonality, the carrier spacing would be 250 Hz. Thus the total bandwidth required is now 250 kHz. There is thus a simple relationship between the bit rate, the bits per symbol, the symbol rate, the carrier spacing, and the required channel bandwidth.

At this stage many readers will rightly regard COFDM as quite complex, and it is that complexity that has delayed its commercial application for over two decades. Soon, we will show how modern digital signal processing has come to the rescue, but first it will be worthwhile outlining the advantages of COFDM that have made tackling the complexity worthwhile.

Properties of COFDM

One of the outstanding characteristics of COFDM is its resistance to multipath propagation. First, let us look at this from the point of view of the frequency domain. A frequency sweep applied to a channel showing multipath effects would normally look a little like Figure 9.2, which shows the path loss plotted against frequency. There is normally a dominant echo signal on such paths which, in the frequency domain runs into and out of phase with the main signal. Consequently, the frequency sweep shows broad peaks where the echo is in phase and narrow but relatively deep minima where the echo is out of phase.

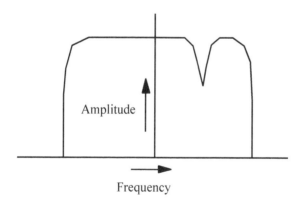

Amplitude

Frequency

Figure 9.2: Channel Frequency Response with Multipath

With analog modulation, these dips can cause signal distortion by removing a significant part of the energy in the sidebands. More particularly, if they happen to attenuate the carrier, they may greatly disrupt the process of demodulation, with results that listeners to broadcast FM channels in cars have often experienced.

Studies of multipath propagation in urban and suburban situations have shown that the frequency ranges, over which the dips in the frequency response are considered deep, span from a few hundred hertz to a small number of kilohertz. If one could arrange a transmission system so that energy taken out of the signal over these ranges always caused little distortion, then the effects of multipath would largely be overcome.

The COFDM carrier spacing and bandwidth can be arranged so that the dips in the frequency response of typical propagation paths severely attenuate, or may take out completely, only a few of the possibly many hundreds of carriers that may be being transmitted in the channel. Since the bit-stream is spread across all the carriers, some carriers going missing means that only a few bits are absent or are wrongly demodulated in the delivered bit-stream. Furthermore, it is common to distribute the bits in a pseudo-random manner across the carriers. Thus, a missing carrier gives rise to an essentially random, rather than a periodic, pattern of missing bits. A reasonably robust forward error correction scheme can accurately correct or replace the missing bits in most circumstances. This scattering of the bit stream across carriers accounts for the term "coded" in the title.

Now let us look at the effects of multipath in the time domain. In urban and suburban environments, the echo signals in most propagation paths are found to arrive within a few tens of microseconds of the main signal. Hence, if we could arrange the symbol period of the digital signals on any carrier to be much longer than this, the echoes from earlier symbols would have died out within a few percent of the duration of any particular symbol. This argues for long symbol durations and slow consequent signaling rates, which is of course just the situation that applies to each carrier in the COFDM channel. In fact, all applications of COFDM apply a "guard band" that is a short time over which the receiver ignores the demodulator output at the beginning of each symbol period, allowing any path echoes from previous symbols to die out before the receiver tries to estimate the signal values. Typically, guard intervals might encompass 10% to 25% of the symbol duration.

After the guard interval, of course, the spurious multipath signals cause a symbol to suffer from interference from delayed echoes of itself. However, it is easy to visualize the effects of this as causing a constant error in phase or amplitude, which once the transients have died out in the guard band is constant from symbol to symbol. Hence, all of the carrier modulation schemes used with COFDM represent symbols by defined changes in amplitude and phase, rather than by using specific fixed values. To a first order approximation, the starting and finishing points are affected equally by interference to themselves outside the guard interval, and hence the change between the two is largely unaffected by the interference. Since the change defines the symbol, the probability of identifying the correct symbol remains high. Transmitting the changes like this is known as differential code transmission.

Figure 9.3 illustrates the concept. Here, suppose that a carrier phase shift of 90 degrees between one sampling instance and the next represents the bit pair 01. This may happen, for instance, with differential quadrature phase shift keying (QPSK) modulation. In that case, a phase shift of 180 degrees might represent 10 and so on. The two large vectors in Figure 9.3 represent the carrier phases before and after the transmission of the particular bit pair. To these we have added two small vectors representing echoes of the main vectors due to multipath. Notice that, at least to a first order, the echoes do not change the overall phase shift of the carrier between the two sampling instances.

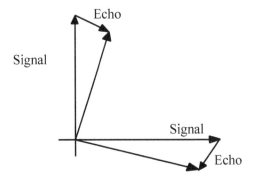

Figure 9.3: Effect of Echo on Differential Phase

Now let us expand on this for echoes from moving reflectors. In this case, the path difference between the main signals and the echoes will vary over the time between the sampling instances, and the phases between the signal and echo at the two instances shown in Figure 9.3 will be different. In this case, the echoes do have a first order effect on the differential phase shift and will act to reduce the noise immunity of the system.

Notice that the change of phase between the main and echo signals over the symbol period constitutes a change in frequency between the two, caused by the fact that the reflector is moving. This is the well-known Doppler shift due to reflections off a moving target. It is also the cause of aircraft flutter in analog TV systems, whereby the echo's running in and out of phase with the main signal causes fluctuations in picture brightness.

Doppler shift is important when COFDM is used for mobile applications, such as Digital Audio Broadcasting (DAB). As we shall see, the DAB standards adopt different operating modes based on the expected incidence of Doppler shifts in different circumstances.

In summary, we see that two very effective techniques combine naturally in COFDM. The use of a great many carriers means that a few missing is readily accommodated. Second, the use of long symbols ensures that transient multipath effects disturb only a small part or their time slots.

It is worth noting that a somewhat incidental, but still very useful, property arises from the multicarrier nature of COFDM. Suppose that a narrow band system is interfering with a COFDM channel. It is easy to arrange matters so that the transmitter suppresses the affected carriers, by setting their amplitudes to zero, and the receiver ignores anything found on their frequencies. Conversely, if a COFDM signal interferes with another service, the offending carriers can be dropped in a like manner. In digital TV applications, for example, removing those COFDM carriers which correspond to the vision, color, and sound subcarriers of the analog TV system has been found most effective in permitting co-channel operation of digital and analog services.

Single Frequency Networks

The capability of forming Single Frequency Networks (SFNs) is one of the most intriguing aspects of COFDM. To see how these networks operate consider a service area surrounded by several transmitters so that a receiver in the area can "see" or receive from some or all of the transmitters. Let us suppose that the transmitters are carrying COFDM, are all operating on the same frequency, and are all carrying the same digital data stream. With conventional modulation systems this would produce intolerable interference. But consider what happens with COFDM.

The receiver sees the strongest signal as the main signal. The other signals on the same frequency and with the same data arrive at slightly different times according to the propagation distance from each transmitter. Thus the receiver overall sees a main signal and several identical smaller versions of it slightly scattered in their times of arrival. But this is just the situation that prevails under multipath conditions from a single transmitter. Hence, all the COFDM properties for defeating multipath effects are called into play as described, and the receiver output is essentially perfect and error free, provided only that the scatter of arrival times remains within the guard band time interval. This is easy to arrange in the design of most networks.

The obvious advantage of SFNs is that we can use three or four geographically spaced transmitters to cover any particular region. Thus, locations that may be shadowed from one will almost certainly be covered by the others. The service area thus becomes very solidly covered with almost no shadow areas from buildings, hills, or trees. However should there be any small location that somehow is not filled, one can feed it with a "gap filler." A gap filler is a simple non-frequency-changing repeater and could consist merely of an amplifier connected between receiving and transmitting antennas with enough isolation. It amplifies and retransmits the off-air signals and, as far as the receiver is concerned, is just another transmitter in the SFN.

In an SFN, the carrier phases and data timings at each transmitter must be locked to reference sources, of course. These references are fairly readily available wherever modern telecommunications links like ISDN are used for signal distribution to the transmitters, but interestingly, the GPS satellite system has also been employed as a timing reference.

Generation of COFDM

Conceptually, the simplest way of generating the COFDM signal would be to use a set of parallel oscillators and mixers as illustrated in Figure 9.4.

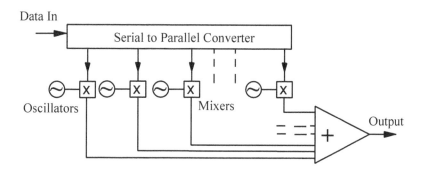

Figure 9.4: COFDM Generation in Principle

Given that there may be hundreds or even thousands of carriers involved in any system, this method is obviously impossible in practice. In fact, although the advantages of COFDM were recognized as far back as 1971, its commercial development did not commence until around 1990. The breakthrough came in the form of powerful digital signal processing with fast integrated circuit chips. The solution is both simple and elegant.

If one writes out the mathematical expression for a group of carriers modulated as in COFDM, it looks remarkably like the expression for Fourier transforms found in most textbooks. In fact one can show that it is equivalent to modulating a single carrier with a signal given by a Fourier transform taken across the data streams on all of the carriers. This is just the process to which high speed digital signal processors are now being applied, and it can be made very rapid by using algorithms like the fast Fourier transform. Hence, the generation of the COFDM signal can now be realized as easily as shown in Figure 9.5. (Strictly, the process involves the inverse Fourier transform, but this is just a mathematical quibble.) The receiver, of course, simply uses the inverse digital transformation. The processor speeds now becoming available mean that this technique can be applied to real time applications like those outlined in the next section.

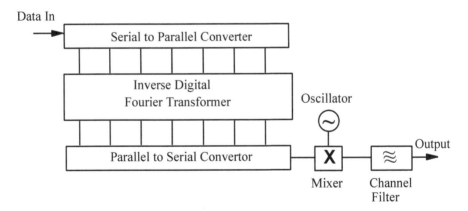

Figure 9.5: COFDM Generation in Practice

Current Applications

Currently, the most developed applications of COFDM lie in the field of broadcasting for sound, (DAB), and for Digital Terrestrial TV Broadcasting (DTTB). International technical standards have recently been agreed for both.

For DAB, commercially operating systems were commissioned in London and Stockholm in 1996. Large experimental systems exist in Canada, Germany, and other European countries. All use the SFN concept and have typically some four to six transmitters.

The DAB standards permit several transmission modes, but typically they operate with several hundred carriers in a 1.5 MHz channel with a bit rate of approximately 2 Mbits/s. With MPEG or AC-3 audio coding, each system can carry from five to fifty separate audio programs with the best quality equal to that available from compact discs.

The initial DTTB applications employed 2000 carriers in an 8 MHz European TV Channel. Recent developments in digital signal processing chips have allowed the number of carriers to be increased to 8000. As this gives a longer symbol period it leads to easier design of SFNs. The systems invariably use MPEG coding and can carry up to five or six simultaneous programs.

Extensive experiments have been carried out in Japan on TV outside broadcasting applications, and equipment is becoming available commercially. Finally, it may be worth noting that the modulation systems used for experimental systems for pay-TV over single-pair phone lines also employ a variant of COFDM.

Chapter 10

Satellite Systems

The area in which digital transmission seems set to have its greatest impact is that of direct-to-home (DTH) television broadcasting from satellites. This field has been restricted in the past by the high cost of transponder capacity and the fact that analog technology permitted the transmission of only one program in each transponder. With digital transmission, as we shall see, up to five or six programs can be carried in each transponder. In addition, the cost of powerful satellites in orbit has fallen dramatically. Hence, we have seen subscription DTH systems starting up in the United States and Europe, and making a huge choice of programs available to their customers. This chapter describes the techniques and design parameters underlying these systems.

System Characteristics

Figure 10.1 shows the overall block diagram of a satellite delivery system. An important point to note about it is that it shares many characteristics with the terrestrial systems we have discussed in previous chapters. Thus the video and audio coding standards and equipment, the formulation of the transport streams, and the forward error correction schemes are all common. Only the actual delivery system is different.

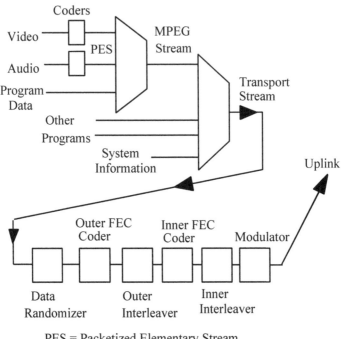

PES = Packetized Elementary Stream
FEC = Forward Error Correction

Figure 10.1: Typical Satellite Configuration

The advantage of all this commonality, of course, is that domestic receivers for the two systems contain many identical components and systems. To change from one method of delivery to another requires little more than a switch from one RF front end and antenna to another. It also means that satellites can very easily be used for program delivery to terrestrial rebroadcast transmitters or cable head-ends with little or no need to decompose the transport streams and bit encoding. As these common elements are all discussed elsewhere in the book, this chapter will focus on the satellite system itself as a digital transmission medium.

Satellite Parameters

The frequency bands used for DTH satellite services have traditionally been based around 3 GHz, known as C-Band, and 12 to 14 GHz, known as Ku-Band. The 3 GHz band has the advantage that the signal is only marginally attenuated by rain, and it has become the preferred option for delivery to rebroadcast transmitters or head-ends. The low frequency and long wavelength dictate that relatively large receiving antennas must be used. These typically range in size from about 3 to 7 meters (10 to 25 ft), which is not a major disadvantage in these applications. However, these antennas are far too large for this band to be used for general DTH purposes.

The 12 GHz band does suffer from rain attenuation, but with modern high-powered satellites, it can provide satisfactory performance in most regions of the world with receiving antenna diameters down to about 600 mm or 2 ft. This is suitable for DTH services, and accordingly the 12 GHz band has become a valuable public resource.

The impending value of the 12 GHz band for broadcasting was foreseen relatively early, and in 1977 it was apportioned to all the nations of the earth by means of an international treaty. The literature on satellite broadcasting often makes mention of this famous conference, referred to universally as WARC'77. WARC stands for World Administrative Radio Conference. A series of these conferences was held under the auspices of the International

Telecommunications Union (ITU), an arm of the United Nations in Geneva.

WARC'77 allocated not only frequencies in the 12 GHz band to each nation, but also orbital slots. An orbital slot is the position a satellite may occupy in the geostationary orbit above the equator. Satellites in the geostationary orbit are at an altitude of about 32,000 km or 20,000 miles above the earth. At this altitude the orbital period is 24 hours, so that as the earth turns the satellite goes around the earth at the same angular rate. Hence, the satellite appears to stay above some fixed line of longitude. This line of longitude denotes the orbital slot allocation. The WARC'77 allocations have been revised to take into account recent developments such as digital technology, but the principle remains the same.

Geostationary satellites appear to remain in a fixed position in the sky, and hence remain within the beamwidth of high gain receiving antennas. Some drift in orbit does occur but is corrected as necessary by on-board thrusters controlled by the satellite operators from the ground.

Transponder Characteristics

In principle, the function that the satellite has to perform is disarmingly simple. It needs only to receive the uplink signal, change its frequency, amplify it, and retransmit it back toward the earth. The equipment that does the frequency translation and amplification is collectively known as a transponder. In practice, of course, the characteristics of transponders are far from simple.

The greatest contributor to complexity in operating transponders is the necessity to squeeze out of them the absolute maximum of transmitter power. Hence, they are conventionally run at power levels where, as amplifiers, they are approaching saturation and consequently exhibiting quite nonlinear transfer characteristics. As such they generate spurious radiation into the adjacent channels served by other transponders on the satellite. The transmitted power limit is nearly always determined by the need to keep this spurious radiation below some acceptable level. The operating power level of the transponder is often described in terms of its "output backoff," quoted as so many decibels. Thus a 3 dB output backoff indicates that the transponder is operating with an output power 3 dB below its saturated level.

The nonlinearity of transponders has generally meant that each could carry only one channel, for if two or more carriers were present the level of intermodulaton products became intolerable at relatively low power levels. Output backoffs of 12 dB or more can be required in these circumstances. Second, analog transmissions invariably used FM modulation as it is very tolerant of nonlinear effects such as amplitude compression in the transponder. Consequently, the bandwidth required to carry a single channel is relatively enormous, being typically 25 to 30 MHz. FM has a high noise immunity, so that effectively its use represented a trade-off of bandwidth for power. Even the best compromise, however, was not particularly good and it seems certain that DTH services would never have been really economic in an analog environment, although some did go into service. Enter digital, however, and the scene has changed enormously.

Digital transmissions through satellites use QPSK modulation as described in Chapter 2. Although we said there that the name is not strictly accurate, the method is close to a pure phase modulation, which implies that the carrier amplitude is relatively constant. A glance at the constellation diagrams in

143

Figure 2.9 shows that the carrier has the same amplitude for all of the constellation positions, whereas this is not true of modulation schemes giving more bits per symbol, such as 16 QAM, for example. Hence, QPSK is by far the method most resistant to the effects of amplitude compression in nonlinear transponders and finds almost universal application in satellite DTH systems.

QPSK provides a transmission capacity of two bits per symbol. The transponder bandwidth on modern satellites is normally in the range of 26 to 72 MHz, with DTH applications tending to favor around 33 MHz. After allowing for band edge effects and the Nyquist frequency rolloff discussed in Chapter 2, one finds that the transmission rate through a typical transponder is about 26 megasymbols per second. The raw transmission rate is thus 52 Mbits/s. The FEC rates for forward error correction can be chosen in the range rate $^1/_2$ to rate $^7/_8$ so that the payload bit rate lies in the range of about 26 to 42 Mbits/s. Given that with MPEG coding a good-quality TV picture requires about 6 Mbits/s, we see that under a digital regime, a satellite transponder can carry from four to six programs depending on the circumstances. This has totally changed the economics of broadcasting from satellites.

Sound Broadcasting

What the favored technology for sound broadcasting may be in the future is far from clear. There are two major competing systems. One is an adaptation of the Eureka 147 DAB standard described in Chapter 12. This uses COFDM modulation that has a far from constant amplitude in the transmission channel. In fact, it exhibits a peak-to-average ratio of around 10 dB. Accordingly, it is very sensitive to amplitude compression in transponders, and these must be run at a considerable output backoff in the satellite. Additionally, a typical ensemble occupies a bandwidth of only about 1.5 MHz so that it makes very inefficient use of the transponders on the current generation of communications satellites. If Eureka 147 is to be broadcast via satellite in the future it will most likely be from purpose-built satellites.

The great advantage of using COFDM would be the capability of using repeaters on the ground to fill in points or regions that may be shielded from the satellite signal by mountains or tall buildings for example. The chapter on COFDM outlines the concept of single-frequency networks, which is readily extended to a satellite service. Thus active repeaters on the ground might receive the satellite signal, amplify it, and rebroadcast it on the same frequency into the shielded area. It would be possible even to radiate the signal into road tunnels and the interiors of buildings by this means.

Despite the problems with transponder nonlinearity, Eureka 147 has been transmitted successfully through communications satellites in some experimental trials, notably in Australia and Brazil. Satellite broadcasting has some obvious attractions for large sparsely settled countries such as these.

The alternative method of sound broadcasting via satellite is being put into service by the Worldspace organization. The aim of Worldspace is to provide an alternative to the international HF radio "short-wave" systems used by such organizations as the Voice of America and the BBC Overseas Service. Hence, one of its imperatives is to be able to serve cheap handheld receivers of a type that would be affordable in the developing countries where these broadcasters find an important part of their audience.

Worldspace uses three satellites operating in L-Band, at around 1.5 GHz, and with the orbital slots spaced evenly around the equator. The audio coding is MPEG-2 Layer 3, to obtain very high degrees of compression such that normal program material requires a bit rate of only 16 kbits/s.

The Worldspace satellites employ two modes of operation. The first caters for those situations where the program sources contributing to the satellite service are widely scattered geographically. Here, it is convenient to permit a number of uplink stations to access the satellite separately, with each having a separate assigned uplink channel.

In this mode, the mode of operation employed in the uplinks is FDMA SCPC, i.e., Frequency Division Multiple Access with a Single Channel per Carrier. Each single channel, with the basic Layer 3 coding at 16 kbits/s, is provided with error correction using an RS(255,233) block code with bit interleaving and a rate one half convolutional code. Thus the overall bit rate is approximately 38.8kbits/s. The modulation is QPSK and the uplink carriers are spaced in frequency by 38kHz.

These uplink streams come together in the satellite and, unusually for a satellite application, are processed within the satellite itself. The processing takes the form of demodulating each stream back to baseband and then assembling the streams into one transport stream at 3.68 Mbits/s. This is modulated onto the downlink carrier, again using QPSK and transmitted with an EIRP of 53 dbW.

The second mode of operation is colloquially referred to as using the transponders in a "bent pipe mode." In this mode all the contributory programs are assembled into the 3.68Mbits/s transport stream at the one uplink station, and transmitted to the satellite on the one QPSK carrier. The satellite transponders merely perform a change of frequency and amplification of the signal before retransmitting it on the downlink.

The first mode of operation offers a simple method of program provision for operators in developing countries where the terrestrial telecommunications may be inadequate, or very expensive, for assembling all the contributory programs at one location. These countries do form a large part of the Worldspace constituency. In general, each satellite transmits two downlinks, one providing this mode of access, and the other accommodating the "bent pipe" application.

The uplinks are in X Band within the frequency range 7025 to 7075 MHz, and the downlinks at L Band in the range 1452 to 1492. The link design parameters permit the receivers to use low gain antennas, usually simple helices. These have low directivity quite wide radiation patterns in the vertical plane. Thus satellite acquisition and tracking is not an issue in the downlinks.

The first satellite in the system, Afristar, was launched in October 1998 and the two others, Asiastar and Ameristar followed in 2001.

Short wave broadcasting is the oldest technology used in the industry. It is interesting that it may well have been the first to be displaced by a combination of the newest technologies in digital transmission and satellites. However, there have been some significant developments culminating as late as April 2001, holding out great promise for the application of digital transmission to short wave services. These developments are described in Chapter 12. There will be much to interest the dispassionate observer as the new satellite and the old short wave proponents come into competition with each other.

Chapter 11

Engineering the Transmission Channel

There are similarities and also some significant differences between the new digital and the old analog requirements in the engineering principles that apply to high-powered transmission systems. Some of the aspects that used to be considered vitally important, such as the very strict requirements on impedance matching in antennas, are now less so. Alternatively, digital transmission brings with it other factors, such as peak voltage clipping in transmitter power amplifiers that were barely considered in the analog world. This chapter is essentially a survey of the differences between the two systems and is intended as a guide for those involved in the design and maintenance of the new high-powered digital systems. The chapter focuses on television, since it is in this area that one finds the subtle differences that those skilled in analog technologies will most usefully come to appreciate. There is very little similarity between digital and analog sound broadcasting technology, at least as world standards now stand. It should be said, however, that the transmission requirements for all digital services are much the same, and much of what is said herein for television also applies to digital sound broadcasting.

Power Requirements

How do digital transmitter powers compare with analog in the same service area? As outlined in the chapter on radio frequency propagation, the answer depends somewhat on the percentages of locations and of time that the system designer may choose to be covered to give a satisfactory result. The maximum ERP permitted will often be mandated by regulatory authorities based on considerations of both coverage and interference to other new or existing services. Thus in the United States, the FCC has determined a table of equivalent radiated power (ERP) values for all the new digital services associated with the current NTSC transmissions.

The discussion of comparative power requirements must take into account the different characteristics of the transmitted waveforms in the digital and analog systems. In NTSC and PAL TV, the peak voltage during the horizontal synch pulses is well defined and constant, whereas the voltage levels during the active picture intervals are not, as they depend on the scene being televised. Hence, the peak synch power has become the accepted way of describing the transmitter power or ERP for these systems. By contrast, in the digital systems the average value of the transmitted signal is well defined but the peaks can be described only statistically, and as we shall see, rare peaks may even be clipped off in the transmitter power amplifiers. Hence, digital transmissions are described in terms of their average power rather than in terms of peak power. When we compare powers for the two, therefore, we are generally comparing digital average power with analog peak synch power.

With what has been said about coverage probabilities and the actions of regulatory bodies, it is probably unwise to generalize about the differential power requirements between digital and analog transmissions into the same service. However, as a rough rule of thumb indicating what might generally be expected, one can say that the digital average ERP will generally be some 10 to 12 dB below the analog peak synch ERP in the same band. This also applies to the transmitter powers if they run into the same or an equivalent antenna system.

Now with the digital services, the transmitter power amplifiers have to be able to handle all but the very rare peak signal levels that may occur. We shall discuss the statistics of these rare peaks later, but for the ATSC 8-VSB system this implies a transmitter power capability of about 7 dB above the average, and for COFDM systems some 10 dB above the average. Hence, overall, the transmitter peak power capability required will lie between being one-half and equal to that of an analog transmitter providing the same coverage. As power supplies, line sizes, and other critical elements in a transmission system are broadly determined by the peak transmitter power, one can expect the overall size and complexity of a digital transmission system to be rather similar to the analog one it might replace.

There is one major exception to this rule, where, as we saw in Chapter 9 on COFDM, single-frequency networks might be adopted. These would replace one single high-powered transmitting station with a network of low-powered transmitters all on the same frequency. This, however, does not seem to be the trend in the initial TV applications, although it is occurring for DAB in Europe. Single-frequency networks are readily realized for COFDM systems, but are yet only at the experimental stage for 8-VSB.

Peak Voltage Levels

Let us look at the question of peak-to-average voltage levels more closely. Recall that in Chapter 2 on the basics of digital transmission systems, we modeled a typical transmitter by passing the data signal in the form of voltage impulses into a filter that defined the symbol shape and channel spectral characteristics. The amplitude of the voltage impulses reflected the particular binary value that applied to each. With real data, these amplitudes are therefore essentially random in that they do not follow any particular pattern. Hence, we have here a random signal exciting a narrow-band filter, and the output can be expected to have some of the character of random noise. Recall also that, except at the receiver sampling instances, the symbol waveforms were allowed to interfere with each other to a considerable extent. Hence, the voltages at any time, other than at a sampling instant, would be the sum over very many symbols leading and trailing in time. It is likely then, that very high peak voltages could occur if some data sequence caused all the voltages from these symbols to be in phase or nearly so. As the data sequences involved would necessarily be long for high peaks, these peaks should be rare, occurring less frequently as the voltage level increases.

For 8-VSB, the voltage impulses in our model can take on any of eight levels. The analogy between the model and COFDM is not so obvious, but in Chapter 9 on COFDM, we saw that it could be generated by multilevel modulation from the output of a Fourier transformer operating on the input data. Thus, COFDM can also be related to our simple model. In TV, the impulse levels can take up to 8000 values in this case. Accordingly, we might expect the transmitted COFDM signal to be even more noise like than 8-VSB. Indeed, this is so. Figure11.1 shows the probability of some level being attained for both 8-VSB and COFDM. The probability is plotted against the level expressed in decibels above the average. These are measured results taken on real systems. On the basis of these measurements the clipping level in transmitters is usually set at between 6 and 7 dB above average for 8-VSB and between 9 and 10 dB for COFDM.

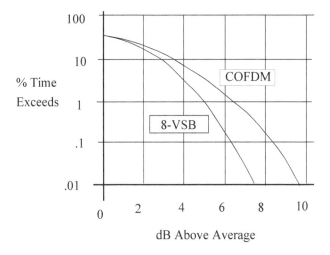

Figure 11.1 Peak-to-Average Probabilities

It is certain that from time to time the input signal will exceed these levels and clipping will occur, although this will be a rare event. When clipping occurs it will most likely cause an error or a short burst of errors. However, as we have seen, all practical digital transmission systems include powerful forward error correction schemes, and hence these rare instances have almost no practical effect as far as the overall quality of service is concerned. This would not be true, of course, if clipping events occurred too frequently. The levels quoted are an accepted compromise between reducing the robustness of the transmissions on the one hand and having to size the transmitters, feeders, and antennas for very high voltages on the other.

There has been some expectation that the peak voltage pulses in digital systems, being brief as well as rare, may not be maintained long enough to cause voltage breakdowns in feeders and antennas. This may have allowed some relaxation in the voltage rating of these components in digital working and permitted them to be run at higher peak powers than would otherwise have been the case. It is found from classical physics, however, that the time constants for voltage breakdowns in air are of the order of 200 nanoseconds. We know from the discussion of symbol shapes in Chapter 2 that the half amplitude duration of isolated pulses is inversely related to the bandwidth of the transmission channel and, for DTV, is around 100 nanoseconds. These two figures are uncomfortably close. The conclusion must be that it is unwise to relax the conventional peak power ratings for digital transmissions.

This argument may not be true for multichannel systems where several transmissions may be carried in the same feeders and antenna. Highly specialized statistical analyses suggest that some relaxation may be permissible, allowing the actual peak voltages to exceed the conventional transmission plant ratings by a few decibels. These studies are at a very early stage, however.

Transmitter Spurious Radiation

In the analog NTSC or PAL systems, the radiated signals contain a small number of dominant carriers, notably at vision, color subcarrier, and sound frequencies. In all digital systems, however, the spectrum is considerably more densely occupied. As would be expected therefore, there is more opportunity for signal mixing to take place because of minor nonlinearities in the transmitter power amplifiers. This is made worse by the process described above whereby the power amplifiers may be allowed to run into saturation at some defined clipping level on peaks. Hence the level of spurious radiation into adjacent channels is likely to be relatively higher in digital transmitters than in analog.

This problem is compounded by the frequency planning and channel allocation policies that will necessarily apply during the period when digital and analog services must coexist until the latter are eventually closed down. To secure enough channels for this period the regulatory authorities are assigning services into adjacent channels in the same service area. In a purely analog situation the adjacent channels were considered "taboo" because of the propensity of analog services to interfere with each other. With mixed services, this adjacent channel working becomes permissible if the relative levels are chosen carefully. It is compromised, however, when the transmitters emit significant levels of spurious radiation into the channels adjacent to those to which they are operating. Accordingly, all regulatory authorities have assigned very severe limits on the permitted levels of spurious radiation from digital transmitters.

The mask adopted by the FCC in the United States is shown in Figure 11.2. This has proven to be a very severe requirement, and most transmitters will not comply without some external filtering on the output. The necessary filters are coming to be called clean-up filters. Simple straight-through filters are also proving to be inadequate. These work by reflecting the spurious power back towards the power amplifiers, which in consequence see a mismatched load just outside their operating channel. A solution lies in using constant impedance balanced ring circuits to contain the filters. Those circuits are the heart of adjacent channel combining, and a description of how they work will be found in Chapter 13 on the mixed digital and analog environment.

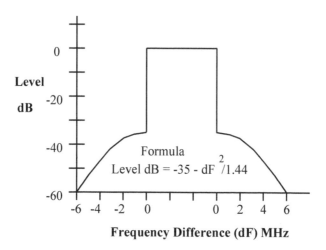

Figure 11.2: FCC Spurious Radiation Mask

Impedance Matching

One of the most difficult, and ultimately one of the most successful, tasks facing the developers of the analog transmission systems was the extent to which impedance matching had to be perfected in high-powered components such as feeders, antennas, and switching frames. This arose because mismatches gave rise to transmitted echoes that appeared on the screen as trailing ghosts. The echoes had to be sufficiently small to make the ghosts invisible to ordinary viewers. Thus antenna return losses of better than 30 dB were the order of the day.

With digital, the situation is quite different. In Chapter 3 on radio frequency propagation, we showed that all digital systems must be designed to be largely immune to the effects of the multipath propagation found in typical urban environments. Accordingly, the ATSC system includes an adaptive equalizer to cancel multipath echoes, and the COFDM systems employ a variety of techniques, including guard bands. Now, there is nothing to distinguish an echo due to an impedance mismatch from one due to multipath propagation, except that the former may be simpler because it is static in time and constant in level. It follows that digital systems have a very high immunity against mismatch echoes, and specifications for impedance matching can be considerably relaxed from those prevailing in the analog era.

One cannot take this too far, of course, because these echoes will absorb some of the systems' capacity to counter echoes arising in the external propagation path. However, it does follow that the specifications applying to transmission plant items designed for the analog service will be more than adequate for digital operation.

System Testing

The very essence of analog transmission systems is that they should faithfully reproduce all sorts of waveforms. Hence a large component of testing an analog system consists in sending standardized waveforms through the system and measuring the difference, or distortion component, by which the reproduced version differs from the original. Thus, in analog TV, we had test signals comprising square waves, luminance pulses, chrominance pulses, and the like.

We saw in Chapter 2 on the introduction to digital systems that the function of a digital transmission system is also to reproduce faithfully one particular signal, that comprising the symbol waveform for the system. Conceptually then, one might expect the digital system to be tested with a test waveform much as analog systems are. However, the only points at which distortion on this waveform is important is at the receiver sampling instances; at other times the distortion is quite incidental to the operation of the system. Now, as described in Chapter 2, any distortion affecting the signal at the sampling instances will cause a spreading of points on the constellation diagram as shown in Figure 11.3 for a QPSK signal.

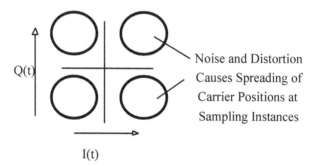

Figure 11.3: Noise and Distortion on Constellation Diagram

Figure 11.3 leads directly to a method of measuring the distortion a system may be producing. If we transmit a random data stream through the system, it is possible to determine the average distance by which all the observed constellation points differ from their correct positions. This is known as the error vector magnitude (EVM). The reason for the name will be obvious from the diagram in Figure 11.4, which illustrates the concept. In effect, for any received signal, we can draw a small vector on the constellation diagram from where the signal should lie to where it actually does lie. Now do that enough times and one can arrive at a statistically significant average for the length of all these small error vectors. When compared to the average length of the signal vectors leading to each constellation point, this gives a meaningful measure of the system's degradation.

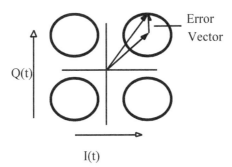

Figure 11.4: Concept of the Error Vector Magnitude

The error vector magnitude is likely to become the single most important measure of transmitter system performance. Specialized monitoring sets known as vector signal analyzers are becoming available and can continually monitor the transmitter while it is carrying program. Naturally, however, the general tests applicable to any radio system, such as the measurement of frequency stability and spurious radiation, will still be required.

We should make one point here about the use of vector signal analyzers with 8-VSB transmitters. An operator might expect to see a constellation diagram like that in Figure 11.5.a since, after all, there are only eight permitted signal levels. In fact, however, the instrument will show a diagram like that in Figure 11.5.b. The former is applicable to straight linear amplitude modulation, but the conversion of the modulated signal to VSB introduces a component on the quadrature modulation axis. The reason for this was described in Chapter 2. In this case, the quadrature component does not carry any useful data, and its presence serves only to preserve the VSB nature of the signal.

(a) Small Quadrature Component

(b) Normal Quadrature Component

Figure 11.5: Constellation Diagrams for 8-VSB

The quadrature component does not have the property of being zero at successive sampling instances. This is not important as the quadrature component does not carry data, and intersymbol interference within it is of no consequence. However, it does mean that its amplitude at the sampling instances does vary randomly depending on the preceding data sequence. Thus the possible constellation positions for each of the eight transmission levels become straight lines rather than simple points and the diagram becomes that shown in Figure 11.5.b.

Using Constellation Diagrams

Whilst the error vector magnitude does provide a very useful quantitative measure of system performance, one can also

160

obtain a good qualitative indication of what may be going on by merely looking at the constellation diagram. All the useful vector signal analyzers present on-screen displays of the constellation diagram of the system under test. Therefore, let us look at two typical displays that an operator might see on one of these instruments.

Figure 11.6.a shows the constellation diagram for an 8-VSB system as it might appear on a vector signal analyzer. This is simulated for a system operating with a demodulated carrier phase error of five degrees and with a signal to noise ratio of about 25 dB. Note that the constellations appear as dotted rather than full lines as the display includes a finite and not an infinite number of sampled points.

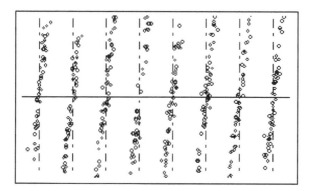

Figure 11.6.a: 8-VSB Constellation Diagram

On this diagram, the noise and the demodulator phase error are easily seen. The diagram also shows something else—that phase error is critical and the system is about to fail catastrophically with a very large error rate. We can see this more easily by collapsing the quadrature display, so that the diagram becomes as shown in Figure 11.6.b.

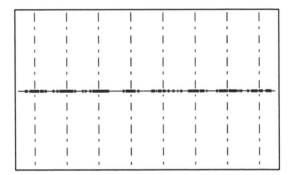

Figure 11.6.b: Collapsed Quadrature Axis

Under this condition, all the sampling points are brought together onto the in-phase signal axis. If they overlap, the receiver will misinterpret the levels being received on that signal axis. Since that component carries the data, errors are bound to occur. Hence, the constellation diagram readily shows the detriment caused by a combination of demodulator phase error and additive noise.

Figure 11.7 shows a constellation diagram for 64 QAM under the same conditions of demodulator phase error and noise. Similarly to before, visualize either the in-phase or the quadrature signal displays being collapsed. It is evident then that the sample points are about to overlap on either axis. Hence this system too is about to run into a high error rate.

Notice that the constellation diagrams for both these systems have shown how critical demodulator phase errors can be. Therefore in system designs, the circuitry that demodulates the phase reference is invariably quite elaborate and complex. They generally take the form of phase-locked loop circuits. The manner in which these operate is essentially as explained in Chapter 4 for the extraction of synch bites in transport streams.

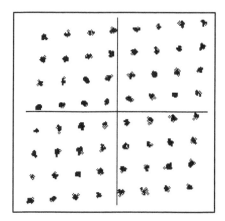

Figure 11.7: 64QAM Constellation

Eye Diagrams

In a chapter dealing with the testing of digital systems, it would be remiss not to mention eye diagrams, as they were once regarded as the unassailable method of choice. However, they now do not find much application in broadcasting, for reasons that will become clear.

The simplest way of describing eye diagrams is probably to outline the way they are generated and viewed. Imagine, therefore, that we have a storage oscilloscope connected to the output of receiver filter, immediately before the digital sampler. The oscilloscope sweep is locked to the oscillator driving the sampling gate in the receiver. The sweep length is adjusted to encompass about three to five symbol periods on the incoming data. The storage time is long enough to hold on the screen some few hundred successive sweeps superimposed on each other.

If the data stream is reasonably random, the oscilloscope will come to show many voltage versus time "paths" that can occur in the receiver at the input to the sampling gate. With enough sweeps superimposed, the display will include nearly every possible path. As an example Figure 11.8 shows such a diagram for an idealized and unimpaired three-level system.

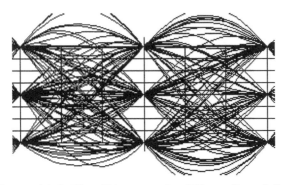

Figure 11.8: Eye Diagram for Three-Level System

The horizontal sweep duration here encompasses just over two symbol periods. The receiver sampling instances occur in the middle of the diagram and adjacent to each outer edge. Notice that here all paths pass accurately through the three correct signal levels and there would be no ambiguity in the receiver as to which is the one transmitted. The openings around these points in the diagram constitute the "eyes."

Where there is some impairment in the system, these paths deviate more or less from the correct levels. Each received symbol waveform may be distorted and not pass through zero at the sampling instances of all the others. This is intersymbol interference. As many symbols contribute to it at each sampling instance, the deviation from the correct levels differs for each possible transmitted data sequence.

Figure 11.9 shows the eye diagram for an 8-VSB signal with the same impairments as used previously, namely a five degree demodulator phase error and about 25 dB signal to noise ratio. Both the phase error and the noise contribute to the filling-in, or closure, of the eyes.

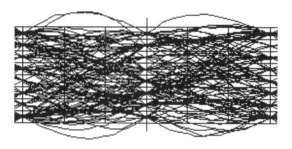

Figure 11.9: 8-VSB with Phase Error

Notice how much more difficult this diagram is to interpret than was the constellation diagram. The principal reason for this is that as the number of symbol levels increase, the eyes become a vanishingly small component of the diagram. At even eight levels the diagrams are of doubtful usefulness. This is the reason for the statement that they will not often be seen in broadcasting applications.

Chapter 12

Transmission Systems in Practice

Throughout the book we have looked at a number of the characteristics of practical digital broadcasting systems to illustrate the principles that underlie digital transmission. This chapter brings together in a concise form the features of the principal systems now coming into service around the world.

It must be said, however, that there are a great many details of these systems and that it would be impractical to cover all these in one chapter, or perhaps even in one book. These are available in the applicable standards where the interested reader will find that the description of even seemingly simple matters like the structure of the transport streams takes up a dauntingly large number of pages. The intention of this chapter, and for the whole book, is to give the reader an overview of the principles involved and an appreciation of how they have been put into practice.

In fact, only a very few systems can really be regarded as world standards, and as we shall see, these share many features. Hence, we will commence the discussion by describing the typical configuration found in all of these systems.

Typical Configuration

Figure 12.1 shows the typical transmitter end configuration for a digital broadcasting system. Here we show encoding of video and audio in a multichannel configuration. Of course this could well be for only one channel or for audio only in the simplest applications.

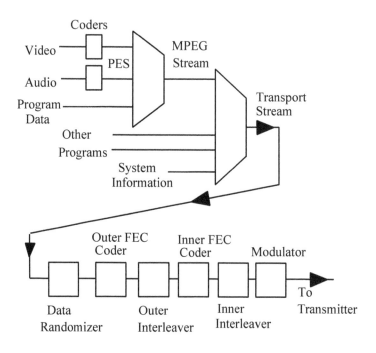

Figure 12.1: Typical Transmitter Configuration

The starting points for our universal transmission system are the coders that compress the incoming video and audio signals, giving the packetized elementary streams (PESs) that we discussed in Chapter 4. These streams are combined in a multiplexer with program associated data to give the MPEG transport stream. The transport streams from a number of sources may enter a second multiplexer together with any data streams and a new service channel we have not previously discussed, known as the system information (SI) channel.

The SI is essential in a multichannel system. It contains information on all the programs and data streams being carried. In particular, it tells the receiver what packet ID (PID) code is associated with each service. Thus, if a receiver is to be "tuned," to use the word loosely, to, say, program No. 2, it will decode only those transport packets carrying the appropriate PID and discard the rest. The SI carries other essential system information as well. For example, where the modulation parameters can be reconfigured, as in those systems employing COFDM, it can inform the receiver of impending changes to ensure that it alters its configuration at the correct instant. The SI is often transmitted with a more robust modulation method than the program signals themselves. For example, a smaller number of levels may be used in the digital symbol waveform while it is being transmitted. This is done to ensure that the SI is more immune than are the programs to interruption by adverse events such as noise bursts. Thus the receiver retains its correct settings and timings, and the program recovery is rapid after such an event.

The multiplexer essentially combines all the input signals into the transport stream. The data stream from it feeds into the data randomizer, sometimes called the energy dispersal unit. Its purpose is to prevent the transmission of long strings of fixed or repetitive data such as, say, a repeating one-zero sequence or a long continuous run of ones or zeros. When these occur, they tend to make the signal out of the modulator sit on some fixed level or value, such that it resembles an unmodulated carrier. This can cause long-range interference to other services and is best prevented. The randomizer ensures that the modulation resembles a random signal and thus disperses the carrier energy over the transmission channel.

Then comes the outer coding for forward error correction (FEC). This is almost universally a Reed-Solomon cyclic redundancy block code as described in Chapter 5. After that we find the outer interleaver, which, as described in the same chapter, breaks up bursts of errors so that they become scattered single-bit errors in the data stream and are more amenable to correction by the FEC.

Following the outer interleaver, we come to the inner FEC coder, which is universally some type of convolutional or trellis coder. It may be worth mentioning that this inner coding is generally not used on cable distribution systems, as that environment is considered quite benign. Finally, we come to the outer interleaver, which, again as described in Chapter 5, tends to place the bits convolved together by the trellis coding far enough apart in time so that not all are affected by single noise burst.

The system diagrams shown here contain only one set of FEC coders and interleavers. In systems which permit hierarchical modulation, it is possible to have more than one set of both. This permits, in effect, two data streams with different degrees of robustness to be superimposed on the one modulator. Hierarchical modulation is described in Chapter 2 and the manner in which error correction is applied in Chapter 7.

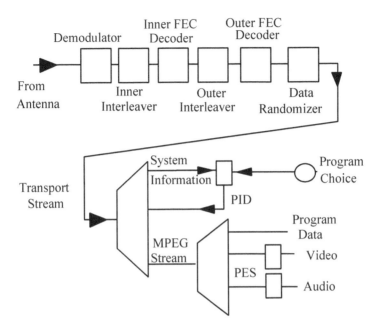

Figure 12.2: Typical Receiver Configuration

Figure 12.2 shows a typical receiver, which is in a sense the inverse of the transmitter diagram in Figure 12.1. Notice here that the program choice determines which PID is sought out by the transport demultiplexer and hence which transport stream packets are sent to the final MPEG demultiplexer.

It is in the modulation scheme that one finds the real differences among the standards that currently exist. The ATSC system developed in the United States uses 8-VSB transmission which is described in detail in Chapter 8. The DVB terrestrial system uses COFDM, as does the Eureka DAB system for sound broadcasting. COFDM is described in Chapter 9. Satellite TV systems universally use QPSK as described in Chapter 10. The following paragraphs list the major characteristics of these systems in turn.

The ATSC 8-VSB System

The ATSC system closely reflects our universal digital system as described. In this case, the video compression system is MPEG-2, and the audio compression is the Dolby Laboratories AC-3. We have already seen many details of the ATSC system as it was used in Chapter 8 to explain many of the principles underlying VSB transmission. Table 12.1 therefore lists only its major capabilities as a transmission system. The figures apply to the system operating in a 6 MHz channel as employed in North America. However, the system can be adapted to channels of different bandwidth by simply scaling the symbol rate.

There is an extension to the standard for 16-VSB modulation for application in cable systems. It operates also in a 6 Mhz channel and twice the bit rate at 38.6 Mbits/s but has a carrier-to-noise (C/N) threshold about 13 dB below that of the eight-level system. This is acceptable because the cable environment is much more benign than is over the air propagation. In fact, some ITU recommendations recognize systems for 2 and 4-level VSB systems as well, but at this stage none seem to have found any practical application.

The term RS(207,187) defining the outer FEC arrangement in Table 12.1 indicates that the FEC is Reed-Solomon in which the blocks comprise 207 bytes, of which 187 are message bytes and 20 are check bytes, as described in Chapter 5 on error correction.

Item	8-VSB	16-VSB	Units
Bandwidth	6.0	6.0	MHz
Symbol Rate	10.76	10.76	megasymbols/s
Outer FEC	R/S(207,187)	R/S(207,187)	
Inner FEC Rate	2/3	1	
Payload Data Rate	19.3	38.6	Mbits/s
C/N Threshold	15.0	28.3	dB

Table 12.1: ATSC System Characteristics

It is interesting to see how the comparative bit rates in these two systems arise. In Chapter 8, we showed how the symbol rate of 10.76 megasymbols per second arises from a channel bandwidth of 6 MHz. The 8-VSB system uses 8-level modulation and therefore can transmit 3 bits per symbol. The 16-VSB system uses 16-level modulation giving four bits per symbol. However, the 16-level system does not employ an outer trellis coder whereas the 8-level system uses a rate two-thirds coder as explained in Chapter 8. Dividing the four-thirds factor for the extra levels by the two-thirds factor for FEC, we see that the 16-level system should have twice the bitrate capacity of the 8-level system, as indeed Table 12.1 shows it does. However, the price is a loss of over 13 dB in noise immunity.

DVB Terrestrial System

This system uses COFDM modulation as described in Chapter 9. Both video and audio compression are MPEG. It is not so easily described as the ATSC system, as the standard permits a great deal of variation. Most important, there are two choices for the number of carriers in the COFDM ensemble, giving what is known as the 2K or 8K modes. The number of carriers in the 2K mode is 1705, and in 8K mode it is 6817. Notice that this is nearly 2000 and nearly 8000, respectively, and the reason for the nomenclature 2K and 8K may be obvious.

Table 12.2 gives some critical values for what are regarded as fairly normal configurations.

	8000 Carrier Mode	2000 Carrier Mode
No. of Carriers	6817	1705
Carrier Spacing	1.1 kHz	4.5 kHz
Symbol Duration	896 usec	224 usec

Typical Data Rates

	QPSK	16 QAM	64 QAM
FEC = 1/2	5.9	11.7	17.6
FEC = 2/3	7.8	15.6	23.4
FEC = 3/4	8.8	17.6	26.4

Mbits/s with 1/16 Guard Interval

Table 12.2: DVB Terrestrial System Characteristics

As Table 12.2 indicates, the modulation of the carriers in the COFDM ensemble may be changed between QPSK, 16 QAM and 64 QAM. The forward error correction may also be varied from rate $^1/_2$ to rate $^7/_8$, and finally, the guard interval may range over several values from $^1/_{32}$ to $^1/_4$ of the symbol interval. Hence, the number of possible permutations is enormous. This applies to European 8 MHz channels, and the standard permits direct scaling for other bandwidths.

DVB Satellite System

As outlined in Chapter 12, the DVB standard for satellite transmission specifies QPSK for the modulation system. It does, however, permit several values for the FEC rates. Table 12.3 shows the available data rates and required carrier-to-noise ratios for a typical transponder bandwidth of 33 MHz. The bit rates can be scaled for different transponder bandwidths.

FEC Rate	Payload (Mbits/s)	C/N Reqd.
1/2	23.8	4.1 dB
2/3	31.7	5.8 dB
3/4	35.6	6.8 dB

For 33 MHz Transponder Bandwidth

Table 12.3: DVB Satellite System Characteristics

Eureka 147 DAB System

The Eureka 147 for digital audio broadcasting has essentially the same block diagram as that shown for the universal digital broadcasting system and lacks only the video encoder. Of course the bit rates are quite a lot lower than in the TV transmission systems and the required channel bandwidth is only

1.5 MHz. The modulation system is again COFDM. The standard describes four modes of operation, with the characteristics as shown in Table 12.4. The four modes arise because the relative immunities to multipath echoes, Doppler shift, and local oscillator instability depend on the symbol durations and carrier spacings in different ways. A broadcaster can choose the mode that best suits his circumstances.

	Mode 1	**Mode 2**	**Mode 3**	**Mode 4**	
Max Transmit Freq.	300	1500	3000	1500	MHz
No. of Carriers	1536	384	192	768	
Carrier Spacing	1	4	8	2	kHz
Symbol Duration	1246	312	156	623	usec
Guard Interval	246	62	31	123	usec

Table 12.4: Eureka 147 DAB Characteristics

The DRM/iBiquity IBOC System

In early 2001, the in-band on-channel audio broadcasting system being proposed by the Digital Radio Mondiale (DRM) consortium, and for which the iBiquity group is the main proponent in the United States, achieved the status of both a world and a European Standard. It has been adopted by the ITU as ITU-R BS.1514 and by the European Telecommunications Standard Institute as ETSI TS 101 980.

The DRM/iBiquity system is very flexible in permitting a number of modes of operation, depending on its particular application. At present, the standards apply only in frequency bands below 30 MHz, that is to MF and HF broadcasting, but it seems very likely that the system, or something very like it, will soon be approved for use in the FM broadcasting band.

The audio coding is MPEG-2 AAC, or a similar system called Perceptual Audio Coding by iBiquity in the United States. The transmitted bit rate is 48 kbits/s for stereo or 24 kbits/s for mono audio in the AAC version. The transmission method is COFDM with approximately 200 carriers in a bandwidth of 9 or 10 kHz. The carriers are modulated with either 64 or 16 QAM, depending essentially on whether a sky wave (HF) or ground wave service (MF) is intended. A particularly robust form of interleaving, as described in Chapter 5 on Forward Error Correction, is used. In the HF application, the interleaving can be as long as 2.4 seconds to combat the signal fading that commonly affects these services. The recommended interleaving time for MF services is 0.8 seconds.

Satellite Sound Broadcasting Systems

There are effectively two systems recognized by the ITU for sound broadcasting from satellites. One, known as Digital System A, corresponds simply to transmitting the Eureka 147 ensemble as described above through a satellite. The other, known as Digital System B, is essentially the same as the Worldspace system described in Chapter 10. It is said that System A should be preferred in the frequency bands below 3 GHz, and System B for frequencies above that. This, however, seems to be an oversimplification, and the choice will depend on many factors, including, in particular, the affordability of receivers to the target audience.

Chapter 13

The Mixed Analog and Digital Environment

Although digital transmission technology has developed apace in recent years, one can be certain that it will not replace the existing analog system overnight. The viewing public has a great deal of money invested in the population of existing receivers. Many are not going to be easily persuaded to change over to new and potentially far more expensive sets, even if they can be convinced that the quality of what they are hearing and seeing is going to be enhanced. It is doubtful that improved quality alone is a sufficient driving force for a rapid public acceptance of new technology. Many may remember how long it took for color TV sets to displace the population of black and white models after the introduction of color TV transmissions—and color offered a major advance in perceived quality. Hence, we can expect that it will be a number of years before the analog network can be closed down, and during that time the digital and analog transmissions must coexist. In fact, those governments that have announced formal timetables for the introduction of digital TV are allowing generally from six to eight years for the changeover. Almost always there is the caveat that this is to be reviewed, suggesting some lack of confidence that even that timetable can be achieved.

This chapter examines the factors that apply to a mixed digital and analog environment. It looks at how transmission channels are being made available to fit the new services in with the analog services currently operating. There are issues of matching the analog and digital coverage areas and of controlling mutual interference between the two. The chapter addresses the use of external plant carrying mixed analog and digital signals. It makes the assumption that the broadcasting organizations currently running analog services will want to move into the digital domain and use the existing infrastructure, such as antennas, towers, and feeders, to the maximum possible extent.

The treatment is largely confined to television broadcasting. As noted in Chapter 12 in the discussion of practical digital systems, there has been until recently only one standardized system for sound broadcasting; the Eureka 147 DAB system. Generally speaking it is not used in the same frequency bands as the existing analog systems, and so the question of how the new and old systems might operate together and share facilities and frequencies does not really arise. It may in the future, when DAB enroaches into Band II, currently occupied by FM services. However, it is generally envisaged that DAB will develop in other bands until it has reached the degree of acceptance and penetration at which the FM services may be closed down. However, some countries are allocating vacant TV channels at the top of Band III for DAB services right from the outset. Generally speaking, however, the transmitter powers are so much lower than for TV and the transmitted bandwidth so much less that the sharing of any significant infrastructure will seldom be necessary.

Once the in band on channel (IBOC) or in band adjacent channel (IBAC) systems mentioned in Chapter 1 are developed, then we will have a very intensive sharing of frequencies and plant between digital and analog sound services. It is probably true to say that devising successful methods of sharing frequencies with the analog services has been the difficulty causing most delay in the realization of a fully accepted IBOC or IBAC scheme. For the moment, however, let us look at the situation with TV, where there is much to interest those faced with embracing the new technology.

Channel Allocation

Historically, a number of constraints have applied to the allocation of analog channels in the same or adjacent service area. These have arisen from the need to avoid mutual interference between the services, which are manifested as artifacts visible to the viewer. Thus, for example, no channel could be allocated if its frequency range included the local oscillator frequency of receivers tuned to another channel in the service area. In this case, the interference generated by those local oscillators could appear as herringbone patterns on the screens of the receivers affected. In particular, adjacent channels, either one up or one down from another channel in the area were very strictly avoided. Beats between the carriers of these channels, produced in receivers, fall into the video band and can produce very annoying patterns on the screen. In general, the forbidden allocations were known as "taboo" channels.

With digital, much of this has changed rather dramatically. Principally, the digital channels carry no strong single carriers with their propensity to produce interference and beat patterns. In analog, of course, we have the vision and sound carriers and the color subcarriers that are always present at high levels. Rather, the spectrum in the digital channels is quite diffuse, with the energy spread very evenly throughout the band of frequencies occupied by

181

the transmission channel. Hence, digital signals are much less likely than analog to cause visible interference. At the same time, as we shall see, they can be made quite resistant to interference from analog channels. When interference does occur to a digital channel, its effect is to lower the signal-to-noise immunity of the system, rather than to produce a visible artifact. Hence, even moderate levels of interference may not be visible at all to the viewer.

Taken together, these characteristics permit channel allocations in a mixed digital and analog environment that would never have been contemplated in a totally analog environment. In particular, they have made it possible to support both analog and digital services in that block of spectrum, the broadcasting bands, that previously supported only the analog channels. Or should we say, they have made it possible to establish the digital services without seeking an additional allocation of spectrum, which would have been exorbitantly expensive or more likely, not forthcoming.

Most of the old taboo rules are no longer applicable. One of the more important is that of allocating adjacent channels. To make the discussion concrete, let us take a service area in which say channels 30 and 33 were allocated. Neither channels 31 or 32 could then have been used for analog services, but either or both can now be used for digital services. The same, of course, applies to channels 29 and 34. In these circumstances the major constraint would probably be the question of what channels are operating in the adjacent service areas. This raises the matter of what protection ratios are required between digital and analog channels in these circumstances.

The protection ratios that authorities around the world are likely to adopt depend very much on local circumstances. Not only do the standards used for both the digital and analog systems differ, but so do the definitions of what constitutes harmful interference. Some authorities use the concept of "threshold of visibility" (TOV), whereas others adopt a certain error rate in the digital channel. Nevertheless, we can generalize to some extent and the following list is intended as a guide.

The following shows the desired-to-undesired (D/U) ratio in decibels between the wanted (desired) service and the interfering (unwanted) service. Where this is negative it indicates that the level of the interfering signal can actually be higher than that of the wanted signal:

Interference Digital to Digital
      Co-channel—D/U of 15 to 20 dB
      Adjacent Channel— D/U of –40 to –45 dB
Interference Analog to Digital
      Co-channel—D/U of 3 to 6 dB
      Adjacent Channel – D/U of –25 to –30 dB
Interference Digital to Analog
      Co-channel—D/U of 35 to 40 dB
      Upper Adjacent Channel—D/U of –12 to –15 dB
      Lower Adjacent Channel—D/U of –8 to –12 dB

These are generally limiting values at which artifacts become visible or at which a defined error rate occurs. System designers would generally allow some safety margin beyond these figures.

Some interesting observations can be made. For instance, we mentioned that the spectra of digital signals are relatively diffuse, and in this respect they reflect the properties of random noise. Hence, the D/U ratios for co-channel interference from digital into digital or analog channels roughly reflect the signal-to-noise ratios at which the wanted signals would begin to fail. Notice that D/U ratios for interference from digital into analog are more critical with digital in the upper adjacent channel than when it occupies the lower adjacent channel. This reflects the fact that in all analog systems the sound carriers are situated very close to the upper channel edge in frequency and are accordingly somewhat sensitive to interference from the abutting upper channel. This effect will be critical when we come to discuss adjacent channel combining.

The digital systems do have the means to lessen the effects of analog interference. Thus the ATSC standard for 8-VSB describes an NTSC filter that may be switched in or out. This is a fairly simple comb filter with notches at the frequencies of the NTSC vision, sound, and color subcarriers. It does produce bursts of interference when fast transients occur in the interfering NTSC channel. These have a duration of about 12 DTV symbol periods and are among the reasons why the trellis FEC is spread over 12 symbols. For systems using COFDM, it is possible to leave out or simply not modulate those carriers having frequencies close to the interfering carriers, and instruct the receiver to ignore anything on those frequencies. The D/U ratios quoted reflect the use of these techniques.

Options for Carrying Digital

Given the extent to which the properties outlined make digital channels available, it is highly likely that existing broadcasters will be allocated channels in the same band as their existing analog services, at least in the UHF bands and in Band III. This means that it may well be possible to share the existing infrastructure of antennas, feeders, and towers between the digital and analog services. A broadcaster in this position has essentially three options:

1. to combine both services into the existing or a new antenna;
2. to split the existing antenna into analog and digital halves;
3. to add a digital antenna to the tower.

There is, of course, the fourth option of completely replacing the existing infrastructure or of building a separate tower and antenna system. If one is able to afford this and obtain the necessary environmental clearances, then it is certainly the simplest technically, but the least interesting. Our purpose in this chapter, however, is to look into the factors that govern the more technically challenging solutions that most broadcasters will have to implement. Let us take up first the matter of channel combining and more particularly adjacent channel combining.

Adjacent Channel Combining

This section is headed adjacent channel combining rather than simply channel combining because the interesting problems occur in the former rather than in the latter. If one is able to successfully implement the combining of adjacent channels, then the combining of nonadjacent channels will seem simple by comparison.

Many readers of this book will not be familiar with the concept of channel combining as in some countries it has rarely found any application in the analog era. However, it offers some very attractive solutions for carrying dual digital and analog services during the time up to when the analog services cease. For this reason, Appendix 2 gives a brief outline of the theory of channel combiners and how they are realized in practice. There the reader will find that combiners possess wide-band and narrow-band ports. Typically, the narrow-band port couples one single channel through to the output, whereas the wide-band input provides a transparent connection through to the output for many channels.

The manner in which combining modules are connected to form a combining chain is shown in Figure 13.1. Here, to be concrete, we have shown channels 30, 31, and 32 being combined, but the numbers are arbitrary.

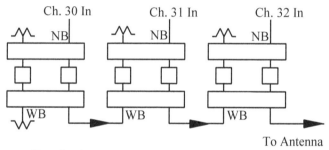

Ch. 30 In   Ch. 31 In   Ch. 32 In

To Antenna

〜   Terminating Loads

NB = Narrow Band Port

WB = Wide Band Port

Fig. 13.1: A Channel Combiner Chain

The important point about channel combiners is that they impose a filter characteristic on the channels that pass through both the narrow-band and broad-band ports. Figure 13.2 shows the characteristics, in terms of amplitude and group delay, that are to be expected in the narrow-band and broad-band ports.

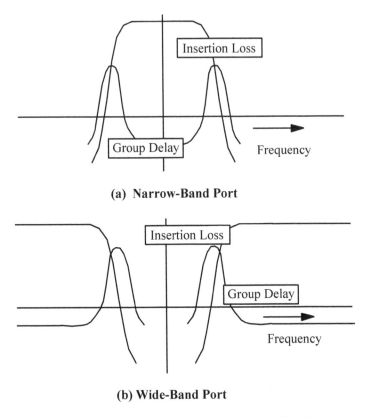

(a) **Narrow-Band Port**

(b) **Wide-Band Port**

Figure 13.2: Filter Characteristics of a Combiner Module

The narrow-band port's attenuation and group delay characteristics accurately reflect those of the filters used in the combiner module, as the discussion of combiner operation in Appendix 2 shows. The wide-band port's response is effectively the inverse of this. There is a small transitional frequency range at each edge of the narrow-band port's passband over which there is some attenuation through both ports. We show in Appendix 2 that there can be no frequency at which the attenuation through both ports is zero, so that these transition regions must exist in all

187

combiners. There is a frequency at which the attenuation through both ports is equal, and it would be 3 dB for both in an ideal combiner made of lossless components. This is often referred as the 3 dB point of the combiner. As we move away from the 3 dB point in the transition region, the attenuation through one port falls and the attenuation through the other rises.

If the combiner is to be used for non-adjacent channels the transition regions cause few problems. They can be placed at frequencies not occupied by the channels being combined. However, with adjacent channel combining, one transition region at least must occur over the frequency at which the channels are contiguous. Hence, both channels will be affected to some extent by attenuation and group delay.

There are two situations to consider here. The first, where the digital signal occupies the channel lower in frequency and the analog channel the upper, causes no particular problems. The second, however, with digital in the upper channel and analog in the lower, needs particular care in implementation.

One of the difficulties that arises is the attenuation of the sound carrier of the analog channel. We mentioned that in all the common analog transmission standards the sound carrier is very close to the channel edge. In fact, where stereo sound is concerned, one carrier may be only a few kilohertz from that edge. Now, if we place the combiner 3 dB attenuation point at the common channel edge, it is evident that there must be some significant attenuation of that sound carrier, since it lies so close to the point at which the port attenuation is 3 dB. The actual attenuation will depend on how steep the attenuation characteristic of the filters in the combiner can be made. Because of the limits imposed by power handling requirements and finite Q values in resonators, eight pole filters are about the steepest that can be used in practice. Hence, one finds that the attenuation of the sound carrier is nearly always 1.5 to 2 dB.

In practice, this means that the power output for the transmitter's sound carrier must be increased. On klystron transmitters this is rarely a problem, but on some solid-state transmitters it may actually be necessary to install additional aural amplifiers.

Of course, by the same argument there must be a similar attenuation of that part of the digital channel that also falls into the combiner transition region. The effects are difficult to analyze, but have been studied by computer simulation. This shows that the amplitude attenuation by itself has a reasonably negligible effect of the digital system performance. The real culprit is the group delay distortion.

In Figure 13.3, we reproduce the spectral characteristics of the ATSC 8-VSB signal. Notice that the carrier pilot falls only 310 kHz above the lower edge of the channel, which puts it well into the region of high group delay for all practical filters. Recall that high group delay indicates phase changing rapidly with frequency. Then, since the system uses the pilot as its phase reference it is not surprising this group delay has a very detrimental effect on system operation.

Figure 13.3: 8-VSB Channel Characteristic

Computer simulations have shown that if the group delay can be corrected, the resulting system performance is quite satisfactory. The most practical method is to place an equalizer in front of the transmitter at baseband. It has been suggested that this could be an adaptive equalizer adopted from receiver practice. In that case, the training signal for the equalizer might be taken at RF from some point in the antenna feeders beyond the overall transmitter output. This realizes a closed loop correction system and in effect, continuously and dynamically corrects all the transmitter characteristics.

When the digital system is one of the COFDM variants, the effect of the group delay is probably negligible, as it will be small compared to the symbol duration on the outer carriers. However, it would be wise to check that this is so in unusual or critical situations.

Antenna Systems

We have previously made the point in this book that there is nothing special required of antenna systems for transmitting digital signals. The digital signals themselves are designed to be particularly robust against the principle problem of ghosting, which is the characteristic of antenna systems of most concern in analog transmission. Hence, it is possible that many existing antennas can be pressed into carrying both digital and analog services during the transition period. One needs to be cautious, however, because the combined bandwidth may be far greater than the original design requirement. Also, where adjacent channels are concerned, there is a requirement that the signal levels in the service area retain their correct ratios. This is to avoid unacceptable interference into the lower-level channel if the relative level of one of the adjacent channels should become much higher than the other.

The two common types of high-powered transmitting antennas are slot arrays and panel arrays. Slot arrays are known to have very narrow impedance bandwidths, inasmuch as an acceptable impedance match to the feeder cables can be achieved over only one or two TV channels. Hence they would be usable only where the digital and analog allocations specify adjacent channels, or sometimes, separation by one vacant channel.

This limitation on slot arrays arises largely because of the nature of the feed system, which is restricted by the small mechanical cross section of the antenna to having the radiating elements connected in sequence or series. By contrast, panel arrays with their greater mechanical cross sections accommodating complex sets of cables can exploit parallel feeding and associated impedance compensation techniques. It seems not often realized that the two different feed techniques also have a major influence on the susceptibility of the radiation pattern to vary with frequency, as we shall now show.

There are three methods of arranging the power distribution to the large number of levels of radiating elements found in high-gain arrays. These comprise:

1. End feeding, whereby the feeder enters the bottom of the antenna and power to the radiators is tapped off the cable as it ascends the antenna column.

2. Center feeding, where the main cable enters the array at its mid-height and splits into one descending and one ascending feeder within the column from which power is tapped off to the radiators in sequence, as before.

3. Parallel feeding, whereby the feed cables split regularly and form a fanlike arrangement spreading over the height of the column. It is readily seen that all the radiating elements are thus fed in parallel, or more importantly, the electrical path lengths to the radiators from the common feeder are all essentially equal.

In end and center feeding, the taps off the feed lines are arranged to be at intervals of one wavelength at the design frequency so that all the radiators are correctly phased at this frequency. As mentioned, end and center feeding are necessarily used in slot arrays to minimize mechanical size, whereas parallel feeding is almost always exploited in broadband panel arrays.

Mechanically, the arrangement of the feed systems in panel arrays is relatively straightforward, as the space allows the use of semi-flexible coaxial feeders formed into cable harnesses. In slot arrays, the slotted external cylinder commonly forms the outer of a transmission line, with the slots being inductively coupled to an internal tubular conductor located coaxially within the outer cylinder and forming the inner of the line. For center feeding, there may be a third conductor inside the lower half of the inner conductor, forming a triaxial line connecting to the common feed point at the center of the antenna. The three different feed systems are shown diagrammatically in Figure 13.4.

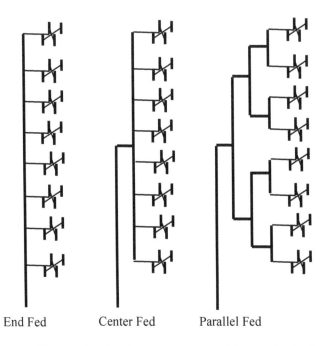

End Fed   Center Fed   Parallel Fed

Figure 13.4: Antenna Array Feed Methods

Vertical Radiation Patterns

In looking at Figure 13.4 it is intuitively obvious that the relative phases to the radiating elements will vary with frequency more in the end-fed arrangement than in the center fed, and this in turn will vary more than in the parallel feed system. In the end-fed array the shortest and longest cables from the common feed point differ in length by an amount equal to the length of the antenna column. In high-gain antennas this may be equal to some 25 to 30 wavelengths. Obviously, the phase difference over this length will change markedly with frequency and cause a frequency-dependent variation of the vertical pattern. By contrast, in the parallel array, the lengths of all the cables from the common feed point are all essentially equal and no relative phase changes are to be expected as the frequency is varied. The center-fed array has characteristics somewhere between the two.

Figure 13.5 shows the extent to which these phase changes can cause the vertical pattern to vary with frequency. Figure 13.5.a is for the end-fed case, 13.5.b for the center-fed case, and 13.5.c for the parallel-fed array. In each case, the design frequency of the antenna is 600 MHz and the plots are taken at this frequency and +/- 6 MHz. The antenna mid band gain is 15 dB; typical of many applications.

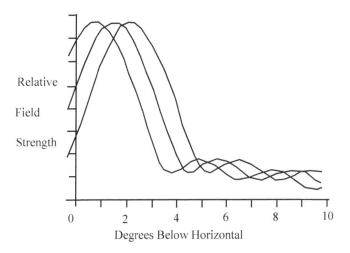

Figure 13.5.a: VP Variation - End-Fed Array

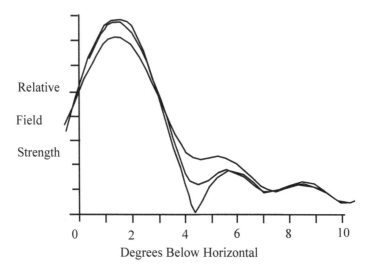

Figure 13.5.b: VP Variation - Center-Fed Array

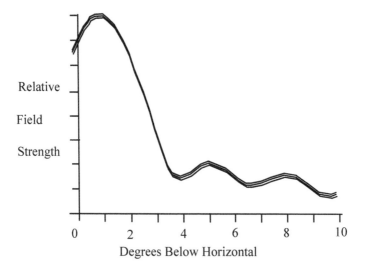

Figure 13.5.c: VP Variation - Parallel-Fed Array

Pulse Distortion in Antennas

Figure 13.5.a showed that with some types of antennas there is a quite marked change in signal levels between adjacent channels. Does this mean that there is some gain slope across each channel? Indeed it does, and this has been long recognized in analog working where it has been considered acceptable in many circumstances. The question arises, however, as to whether it may compromise digital operation more severely. This section reports on some recent work by the author which indicates that severe degradation can occur.

The analysis centers on what may be regarded as the pulse response of the end fed and center fed antennas. Consider a waveform carrying a single pulse, or more precisely a single digital symbol arriving up the feeders and passing into the base of an end fed antenna. The pulse will first start radiating from the bottom radiator. A short time later, as the signal traverses upwards through the antenna, it will commence radiating from the second radiator, then from the third, and so on. In effect, the symbol waveform eventually received in the field is made up of a lot of small replicas of itself, each delayed by some short incremental time. It therefore takes a time, roughly equivalent to that required for light to traverse the antenna aperture, for the pulse to reach full amplitude.

To take a specific example, consider an antenna of 24 wavelength aperture operating at 600 MHz. This is large, but not exceptionally so. The time required for the signal to reach full amplitude by the preceding argument is 40 nanoseconds (nS). Since the half amplitude duration of the digital symbol in the ATSC system is 93 nS, one might well expect that the symbol distortion could be considerable.

The form that the received symbol takes can be calculated and fairly easily simulated by computer. Figure 13.6 shows the result for the 24 wavelength example antenna. The received symbol is superimposed on the ideal waveform (shown dashed), and the distortion can be seen clearly. Notice in particular that the waveform does not go through zero at the sampling instances, and intersymbol interference must therefore result.

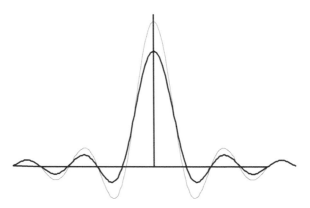

Figure13.6: Symbol from End Fed Antenna

The constellation diagram shown in Figure 13.7 is even more telling. This is calculated at a radiation angle of depression putting it at 0.3 degrees below the center of the main beam. Note that at the sampling instances the in-phase component is quite removed from the correct eight positions represented by the vertical lines. The situation becomes worse further down the flank of the antenna's vertical radiation pattern. Center fed antennas are better, but the distortion is still very significant in most circumstances.

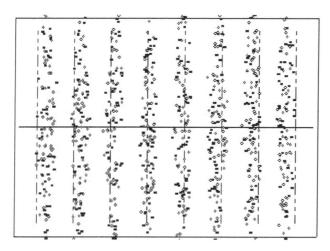

Figure13.7: Constellation from End Fed Antenna

It may be argued that the automatic equalizers in the receivers will correct this distortion. However, calculated quantitative results from calculations show that in many circumstances this action could absorb almost the full capacity of the equalizers. This would surely be very poor engineering practice.

Signal Statistics and Voltage Breakdown

The mixed analog and digital environment is bringing the situation where a large number of services are sharing common items of transmission equipment, like the combiners and antennas referred to earlier in this chapter. In these circumstances, system designers are often surprised at how high the possible peak voltages may become. We find, for example, that several services running at only a few kilowatts each are likely to require the largest co-axial cables, or even waveguide, to handle these voltages, even though the peak power rating of the component is vastly greater than that being applied.

There is probably no better way to illustrate this than by a simple example. Suppose we have five analog and five digital TV services running in one coaxial feeder. Let each analog service have a peak synch power of 1 kW and each digital service have an average power of 1 kW. Thus the total average power in the feeder is around 7.5 kW, depending on how one relates synch and average power in the analog channels. Now we can add together the maximum voltages that each service can attain, remembering to use a peak to average value for the digital services as found in Chapter 11 and Figure 11.1 We find that the peak possible voltage is about 5500 volts. This corresponds to a peak power of about 300 kW! This surely illustrates the point being made.

Whilst these peak voltages can be relatively enormous, it is a matter of simple statistics that they will arise rarely. Most readers will know of what is called *the central limit theorem* of statistics. Simply put, this says that when one adds together many independent quantities, the probability distribution of the resultant approaches *normal* or *Gaussian* in form. The normal or Gaussian probability distribution is very common in statistics. It is often referred to as "the bell curve" in popular scientific literature. Its outstanding characteristic is that increasingly large values have a rapidly diminishing probability of occurring. With the wealth of attention that its application to sums of quantities has received, we can be very confident that it applies to summed voltages in multichannel transmission systems.

To assist with this confidence, Figure 13.8 shows a short extract from a computer simulation of a multichannel system similar to the example just cited. This shows the total voltage in the feeder plotted against time. The y axis tick lines show levels at two dB intervals below the maximum summed voltage levels. We see that the signal virtually never comes within 5 dB of the peak possible value. This suggests that by adopting statistical methods, designers could achieve quite considerable savings on system ratings.

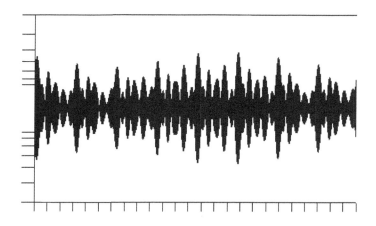

Figure 13.8: Waveform in Multichannel System

In dealing with modulated signals like this, it is more useful to deal with the probability of the *envelope* of the signal, rather than with the signal itself. In Figure 13.8 the envelope is the peak to peak value at any given instance. Notice that this must always be positive—we can never have a negative peak to peak value. The distribution that describes the envelope is the Rayleigh Distribution, which is closely related to the Normal Distribution and can be derived from it. The difference is that, whilst the Normal Distribution is double sided, the Rayleigh Distribution is single sided and applies to quantities having positive values only.

There is yet another refinement we can apply. In the Normal and Rayleigh Distributions the extreme values are not limited and in principle can extend to infinity. This is evidently not the case in a real transmission system. By virtue of each transmitter having some finite output voltage that it cannot exceed, the maximum sum of these voltages cannot also exceed some finite value. Hence the Normal and Rayleigh Distributions must become more and more inaccurate, and pessimistic, in predicting the occurrence of the higher and rarer peaks. The answer lies in the Beta Distribution, which has a finite limit that can never be exceeded. To make this more easily understood, Figure 13.9 shows the three distributions plotted on a linear scale.

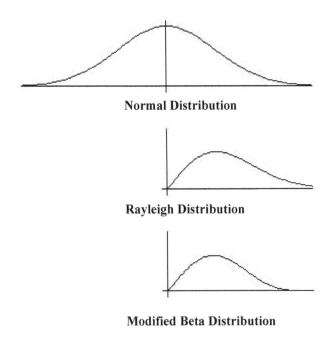

Normal Distribution

Rayleigh Distribution

Modified Beta Distribution

Figure 13.9: Forms of Probability Distributions

When the Beta Distribution is to be applied to the signal envelope, it has to be modified from its usual form. Since the modified version is not generally found in textbooks, I will quote the formula here for those who may wish to pursue the matter further. We have

$$P[E > E_o] = [1 - (E_o/E_p)^2]^{\alpha}$$

where $P[E > E_o]$ is the probability that the envelope E will exceed some given level E_o, E_p is the absolute maximum voltage the system can attain and α has the value $\frac{1}{2}(E_p/E_{av})^2$. E_{av} corresponds to the envelope voltage for the average power being transmitted.

To make use of these statistics we have to take over the concept of a "return period" for very rare events. This is common in civil engineering practice, but generally has been anathema to those working in electrical engineering. The concept relies on the sampling theorem, which was outlined in Chapter 6 in relation to audio sampling rates. Restated, the theorem provides that in a system of bandwidth B, the samples are statistically independent if they are taken at times separated by more than 1/B. B here is the overall system radio frequency bandwidth.

Again, it may be best to explain these concepts by way of a simple example. Taking our ten channel system cited previously, suppose we wish to know how often the peak envelope voltage will exceed a level 3 dB below the absolute attainable peak. From the formula for the Beta Distribution we find that will occur with a probability of one in 10^{-15}. If the system bandwidth is 100 MHz say, then samples taken 10 nS apart are independent. One in 10^{15} will exceed the nominated level and on average that will occur in 10^{16} nS, which is about 3 months. By pursuing this argument, one can find the level that would be exceeded once in one year, once in ten years, and so on.

Some people feel uneasy about the "astronomic" numbers that appear in this argument. Hence, Figure 13.10, which contains a plot of return period versus peak voltage for our example system, may provide some comfort. Note that there is a real cliff effect, and hence the design point of the system is almost independent of the return period chosen. It is almost true to say that there is some voltage that is less than the absolute peak and, for all practical purposes, will never be exceeded. Hence, the application of some simple statistics can provide a substantial benefit in system design.

Figure 13.10: Return Period Vs. Envelope Voltage

The reader should be aware that this statistical approach is quite new and not yet generally applied in the industry. Also, whereas it offers a substantial gain for the rating of multichannel

203

systems, the gain is quite small if only two or three channels are involved.

Nevertheless, voltage breakdown is a subject of active investigation. In addition to the signal statistics we have just discussed, there is the question of whether extremely high voltage spikes might have durations too short to cause breakdown in air. We said in Chapter 11 that this was not so for a single digital channel. In that case, the duration of extreme events roughly matched the time constant applying to the development of an avalanche of free electrons. In multichannel systems, the extreme spike duration is much shorter. Early theoretical and practical work suggests that a relaxation in voltage rating of up to 2 dB may eventually be available from this cause. This may be quite significant in the larger systems.

Appendix 1

The Fourier Transform

In thinking about digital transmission it is very useful to be able to change one's point of view back and forth between the time and the frequency domains. In other words one switches between thinking about waveforms and thinking about spectra. A drawing of a waveform would be a plot of the voltage along a time axis. A drawing of its spectrum would show a certain level plotted along a frequency axis. One can think of this level as representing the energy contained in a very thin slice of frequencies surrounding any given point on the frequency axis. The mathematical process by which one can convert between waveforms and spectra is known as the Fourier transform.

The name derives from that of its inventor Jean Baptiste Fourier, who in 1822 realized that a periodic waveform could be represented by the sum of a number of sine waves with frequencies related to the periodicity of the waveform. The expression giving the amplitude of each sine wave became known as the Fourier series. The Fourier transform is an extension of this. The Fourier series gives a voltage representing the amplitude of discrete sine waves at each frequency involved. The transform on the other hand returns a voltage that, if squared, would give the energy flow around any given frequency. Hence, there is a one-to-one correspondence between the spectrum and the Fourier transform.

The Fourier transforms of many common waveforms have been derived and are available in the literature. Figure A1.1 shows the transform of a rectangular pulse. The transform has the mathematical form of sin(x)/x, referred to as the sinc pulse or more often as the "sine x on x" pulse. The transform of a series of sharp impulses is itself a series of impulses, as shown in Figure A1.2. We have used this relationship in the text to explain the Sampling Theorem, and why high quality sound must be sampled at at least 40 kHz.

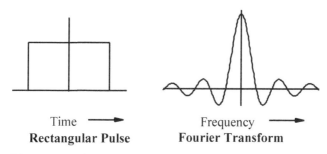

Time ⟶   Frequency ⟶
Rectangular Pulse   **Fourier Transform**

Figure A1.1: Rectangular Pulse and Transform

Figure A1.2: The Sampling Signal and Transform

An interesting thing about Fourier transforms is that they are reciprocal. Thus in Figure A1.3, we show a sinc pulse in time and its spectrum, which turns out to be rectangular in form. Notice that this is just the reverse of a rectangular pulse and its spectrum in Figure A1.1.

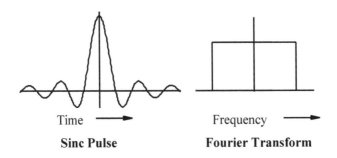

Figure A1.3: Sinc Pulse and Transform

207

Another important point is that a single isolated voltage impulse of essentially zero width in time has a flat spectrum ideally stretching from minus infinity to plus infinity in the frequency domain. Now if we were to pass this pulse through a filter with a rectangular frequency response, the filter would let through only that part of the spectrum that falls within its pass band. In other words the output spectrum would be rectangular, as shown in Figure A1.3, and hence the output pulse must be a sinc pulse, just as shown in that figure. This, as discussed in Chapter 2, is the very basis, the *sine qua non*, of digital transmission.

Appendix 2

Principles of Channel Combiners

As digital TV commences, many existing operators of single analog services may find themselves faced for the first time with channel combining. Even if it is unfamiliar, channel combining may well be the system of choice for running the old analog and the new digital services in parallel during the transition period leading up to a fully digital environment. Where combining is possible, it will almost certainly be cheaper than installing a separate digital antenna with all the implications for tower upgrading. Hence, this appendix is meant to provide a short review of modern combiner principles for those who may not have previously encountered the technique.

Modern channel combiners are based on a circuit known as the constant impedance ring, sometimes called the Lorenz ring after its inventor. As we will show, it has the major advantage of presenting a constant impedance to the transmitter, both inside and outside the transmission channel, and as such minimizes the production of spurious radiation in the transmitter amplifiers. We mentioned in Chapter 11, Engineering the Transmission Channel, that controlling spurious radiation is likely to become a major issue, especially during the transition period when analog and digital services must exist together in limited frequency bands.

Accordingly, not only must the generation of spurious radiation be controlled, but also it is highly likely that most high-powered transmitters will have to be fitted with output filters to achieve the mandatory specification. These filters, called clean-up filters, are also constructed in the form of constant impedance rings. Hence, we can expect these circuits to find wide application in the immediate future.

The 3 dB Directional Coupler

The major circuit component involved in forming the constant impedance ring is the 3 dB directional coupler. Hence, let us begin by reviewing signal flow through these simple devices. Figure A2.1 shows the necessary flow diagrams.

Figure A2.1 shows the relationships between the voltages of the signals at each port when one port is driven and the other three are terminated in matched loads. In this diagram, the use of the letter "j" implies a phase shift of 90 degrees, in accordance with common practice in representing voltages in circuits.

Take diagram (a) in that figure, for example. Here we see that when a voltage V excites port 1, a voltage $V/\sqrt{2}$ emerges from port 2 and one of $jV/\sqrt{2}$ from port 3. Nothing emerges from port 4, and we say that ports 1 and 4 are isolated. Similarly, diagrams (b) through (d) in that figure show the signals at each port when ports 2, 3, and 4 are excited.

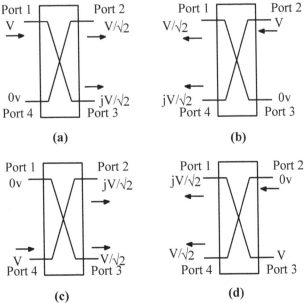

Figure A2.1: Signal Flow in Couplers

Now consider a coupler configured as in Figure A2.2. Here, port 1 is excited by a voltage V, ports 2 and 3 are terminated in open circuits, and port 4 is terminated in a matched load. By virtue of the drive at port 1, signals with voltages V/√2 and jV/√2 emerge from ports 3 and 4. These are immediately reflected by the open circuits and reenter the coupler. Now consider the signal reentering port 2. Because the coupler action is the same as illustrated in Figure A2.1.b and described previously, it must generate a signal of voltage V/2 at port 1 and jV/2 at port 4. Similarly, the signal with voltage jV/√2 reflected at port 3 must produce a voltage jV/2 at port 4 and −V/2 at port 1. This minus sign arises because this particular component has been subjected to two 90-degree phase shifts in its two passages through the coupler. A combination of bold and normal lettering is used on the diagram to assist the reader to trace the component signal flows. Notice now that these component signals add in phase at port 4, but cancel at port 1.

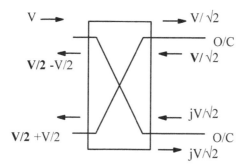

Figure A2.2: Reflections in a Coupler

The ultimate result of the coupler action in this case is to direct all the input power out of port 4. Also, since the net voltage emerging from port 1 is zero, there is no reflected power, and the coupler is matched at this input port.

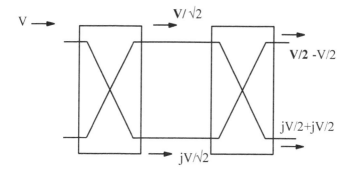

Figure A2.3: Cascaded Couplers

Consider two of the couplers connected together as shown in Figure A2.3. Again we use bold and normal lettering to help with tracing the signal flow. Essentially the signal with voltage $V/\sqrt{2}$ entering port 1 of the right-hand coupler produces an output $V/2$ at port 2, and $jV/2$ at port 4 of that coupler. The signal $jV/\sqrt{2}$ entering port 4 gives rise to voltages $-V/2$ at port 2 and $jV/2$ at port 3. The signals cancel at port 2 and add in phase at port 3. Hence the net result is that all the input power emerges at port 3. All the other ports are isolated from the input signal.

An important question is whether the two lines between the two couplers are matched, since, as we shall see, this could affect the characteristics of the filters to be placed in these lines. To address the question, look at the signal flows into the right-hand coupler. The two inputs are into ports 1 and 4. However, as discussed, these two ports are isolated from each other in normal coupler operation. Hence, no signals emerge from these ports, and none of the power entering is reflected. Hence, by definition, these two ports are matched. We are now in a position to look at how a modern constant impedance combiner operates.

The Constant Impedance Combiner

Figure A2.4 shows the circuit for the constant impedance combiner. It will be obvious why the term *ring* is used to describe this structure. The two filters shown in the ring are identical narrow-band bandpass devices. They define the passband for the port labeled as the narrow-band port on the figure.

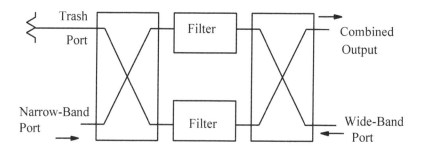

Figure A2.4: The Constant Impedance Combiner

Consider a signal within the passband entering this port. As the filters are effectively transparent to this signal, we have exactly the situation shown in Figure 2.3, and the signal passes through to the output port as explained for that figure. Now consider a signal entering the wide-band port. If it is outside the filters' passband, it will be reflected from the filters, just as it would be by the open or short circuits discussed in relation to Figure 2.2. Accordingly, as explained for that diagram, all of the input power will now emerge from the output port also. Hence, here we have the essentials of the combiner. The signals into the narrow-band port are within the filter bandwidth, the signals into the wide-band port lie outside that bandwidth, and the combiner acts to direct both to a common load.

A few additional points illustrate some of the important properties of these devices. For example, let us suppose the filters exhibit some small mismatch in the passband, as will inevitably be the case with practical devices. This would mean that some signal is reflected back into the output ports of the coupler shown on the left-hand side in Figure A2.4. By tracing the signal flow, as we did for Figure 2.2, it is easy to show that these reflected signals cancel at the input port, but add in phase at the port labeled the trash port in Figure A2.4. Hence, the input remains matched, and this reflected power is directed into the balancing load where it is absorbed. Generally speaking, for a well-set-up combiner, this power will never exceed a few tens of watts in the highest power applications. Similarly, if some signal from the wide-band port happens to leak past the filters, it can easily be shown to finish up in the balancing load also.

We need to qualify this simple picture a little. If the reader cares to trace the signal flows carefully, he or she will find that it is the sum of the two leakage signals that ends up in the balancing load. The difference is directed into the transmitter connected to the narrow-band port and contributes a loss of isolation between transmitters. Similarly, if the in-band reflections from the two filters discussed above are not identical, the difference also is directed back toward the transmitter, causing some mismatch. The primary requirement in designing and setting up these combiners is to match the characteristic of the two filters to the best extent possible.

We have discussed the operation of the Lorenz ring from the point of view of combining, but as mentioned, it also finds much application in transmitter output filters. It will be evident from the discussion so far that the narrow-band port is matched over a broad frequency range, much wider than the operating channel bandwidth. This occurs because all the energy reflected from the internal filters finishes up in the balancing load, whether or not it falls within the passband of the filters. Thus, the transmitter sees a broadband-matched resistive load, as distinct from the highly reactive load it would see from an ordinary filter.

The design and adjustment of the power amplifier are much simpler than they might otherwise be.

The Impossible Combiner

In Chapter 13, on operating combined analog and digital services into one antenna, we stated that some attenuation in each channel is unavoidable. This seems to disturb many people, and the question frequently arises as to whether some clever solution may be found whereby one could realize a perfect combiner with no attenuation in either channel. Some people have spent many hours searching for a realization of this perfect circuit. We shall now show that this ideal is impossible, even in principle.

To show that this is so, we will adopt a form of argument often exploited by mathematicians; that is, we will assume that a solution exists and then show that it leads to ridiculous consequences. Hence, let us suppose that the ideal combiner can be made, and that it will show no attenuation from either input port through to the output. We will call the input ports input 1 and input 2 in this case, since it would not make sense to keep labeling them as narrow band and broadband.

Consider the setup shown in Figure A2.5. Here we have an oscillator with a split output feeding signals with voltage V into each input port. As the combiner ports are matched, the power into each is V^2/R, where R is the characteristic impedance, and hence the total power entering the combiner is $2V^2/R$.

NB = Narrow-Band Port
WB = Wide-Band Port

Figure A2.5: Ideal Combiner Drive Conditions

The voltage at the output due to that at input 1 must be V, since by definition our ideal combiner must have no attenuation. Similarly, the voltage at the output due to that at input 2 must also be V. However, these two signals are at the same frequency, since they are generated in the one oscillator. By adjusting the phase shifter in Figure A2.5, we could bring them into phase, whereby the combined amplitude would be 2V. The power coming out of the combiner would be $4V^2/R$, against the $2V^2/R$ going in. This is clearly impossible, and the conclusion must be that our ideal combiner cannot be realized.

One can refine this argument by assuming that the voltages at the two input ports are unequal in amplitude. The result always shows that if at some frequency the losses from one port are small, then those from the other must be large. Also, if the losses are equal at some frequency, then they cannot be less than 3 dB. This is the basis for the statement made in the book that, for adjacent channel combining, the attenuation at the common channel boundary will be at least 3 dB.

217

Glossary

ADTV	Advanced Digital TV—an early name for the United States Digital System
ATSC	Advanced Television Systems Committee—United States Committee developing standards for DTV
bits/s	Bits per second
CCIR	Initials of the French name for the organization preceding the ITU-R
COFDM	Coded Orthogonal Frequency Division Multiplex
CTV	Cabled TV Distribution System
DAB	Digital Audio Broadcasting
DTTB	Digital Terrestrial Television Broadcasting

DTV	Digital TV
DVB	Digital Video Broadcasting—a consortium, originally European, developing standards for DTV broadcasting
ERP	Equivalent Radiated Power
ETS	European Telecommunications Standard
ETSI	European Telecommunication Standards Institute
Eureka 147	European project to develop DAB
FCC	Federal Communications Commission
Grand Alliance	Consortium of United States manufacturers developing ADTV
ITU	International Telecommunications Union
ITU-R	International Telecommunications Union - Radio
kbits/s	Kilobits per second
Mbits/s	Megabits per second
Megasymbol	One million symbols
NTSC	World standard for analog TV originally developed in the United States

OFDM	Orthogonal Frequency Division Multiplex
PAL	World standard for analog TV originally developed in Europe
Symbol	One isolated digital signaling waveform

Index